Surveys and Tutorials in the Applied Mathematical Sciences

Volume 2

Editors

S.S. Antman, J.E. Marsden, L. Sirovich

T0214935

Surveys and Tutorials in the Applied Mathematical Sciences
Volume 2

Editors

S.S. Antman, J.E. Marsden, L. Sirovich

Mathematics is becoming increasingly interdisciplinary and developing stronger interactions with fields such as biology, the physical sciences, and engineering. The rapid pace and development of the research frontiers has raised the need for new kinds of publications: short, up-to-date, readable tutorials and surveys on topics covering the breadth of the applied mathematical sciences. The volumes in this series are written in a style accessible to researchers, professionals, and graduate students in the sciences and engineering. They can serve as introductions to recent and emerging subject areas and as advanced teaching aids at universities. In particular, this series provides an outlet for material less formally presented and more anticipatory of needs than finished texts or monographs, yet of immediate interest because of the novelty of their treatments of applications, or of the mathematics being developed in the context of exciting applications. The series will often serve as an intermediate stage of publication of material which, through exposure here, will be further developed and refined to appear later in one of Springer's more formal series in applied mathematics.

Surveys and Tutorials in the Applied Mathematical Sciences

1. Chorin/Hald: *Stochastic Tools in Mathematics and Science*
2. Calvetti/Somersalo: *Introduction to Bayesian Scientific Computing: Ten Lectures on Subjective Computing*

Daniela Calvetti   Erkki Somersalo

Introduction to Bayesian Scientific Computing

Ten Lectures on Subjective Computing

 Springer

Daniela Calvetti
Department of Mathematics
Case Western Reserve University
Cleveland OH, 44106–7058
daniela.calvetti@case.edu

Erkki Somersalo
Institute of Mathematics
Helsinki University of Technology
P.O. Box 1100
FI-02015 TKK
Finland
erkki.somersalo@tkk.fi

Editors:

S.S. Antman
Department of Mathematics
and
Institute for Physical Science
   and Technology
University of Maryland
College Park,
MD 20742-4015
USA
ssa@math.umd.edu

J.E. Marsden
Control and Dynamical
   System, 107-81
California Institute
   of Technology
Pasadena, CA 91125
USA
marsden@cds.caltech.edu

L. Sirovich
Laboratory of Applied
   Mathematics
Department of
   Bio-Mathematical Sciences
Mount Sinai School of Medicine
New York, NY 10029-6574
USA
chico@camelot.mssm.edu

Mathematics Subject Classification (2000): 62F15, 65C60, 65F10, 65F22, 65C40, 15A29

Library of Congress Control Number: 2007936617

ISBN 978-0-387-73393-7   e-ISBN 978-0-387-73394-4

Printed on acid-free paper.

Printed in the United States of America. (MVY)

9 8 7 6 5 4 3 2 1

springer.com

Dedicated to the legacy of
Bruno de Finetti and Luigi Pirandello:
Cosí è (se vi pare).

Preface

The book of nature, according to Galilei, is written in the language of mathematics. The nature of mathematics is being exact, and its exactness is underlined by the formalism used by mathematicians to write it. This formalism, characterized by theorems and proofs, and syncopated with occasional lemmas, remarks and corollaries, is so deeply ingrained that mathematicians feel uncomfortable when the pattern is broken, to the point of giving the impression that the attitude of mathematicians towards the way mathematics should be written is almost moralistic. There is a definition often quoted, "A mathematician is a person who proves theorems", and a similar, more alchemistic one, credited to Paul Erdős, but more likely going back to Alfréd Rényi, stating that "A mathematician is a machine that transforms coffee into theorems[1]". Therefore it seems to be the form, not the content, that characterizes mathematics, similarly to what happens in any formal moralistic code wherein form takes precedence over content.

This book is deliberately written in a very different manner, without a single theorem or proof. Since morality has its subjective component, to paraphrase Manuel Vasquez Montalban, we could call it *Ten Immoral Mathematical Recipes*[2]. Does the lack of theorems and proofs mean that the book is more inaccurate than traditional books of mathematics? Or is it possibly just a sign of lack of coffee? This is our first open question.

Exactness is an interesting concept. Italo Calvino, in his *Lezioni Americane*[3], listed exactness as one of the values that he would have wanted to take along to the 21st century. Exactness, for Calvino, meant precise linguistic ex-

---

[1] That said, academic mathematics departments should invest on high quality coffee beans and decent coffee makers, in hope of better theorems. As Paul Turán, a third Hungarian mathematician, remarked, "weak coffee is fit only to produce lemmas".

[2] M. V. Montalban: *Ricette immorali* (orig. *Las recetas inmorales*, 1981), Feltrinelli, 1992.

[3] I. Calvino: *Lezioni Americane*, Oscar Mondadori, 1988.

pression, but in a particular sense. To explain what he meant by exactness, he used a surprising example of exact expression: the poetry of Giacomo Leopardi, with all its ambiguities and suggestive images. According to Calvino, when obsessed with a formal language that is void of ambiguities, one loses the capability of expressing emotions exactly, while by liberating the language and making it vague, one creates space for the most exact of all expressions, poetry. Thus, the exactness of expression is beyond the language. We feel the same way about mathematics.

Mathematics is a wonderful tool to express liberally such concepts as qualitative subjective beliefs, but by trying to formalize too strictly how to express them, we may end up creating beautiful mathematics that has a life of its own, in its own academic environment, but which is completely estranged to what we initially set forth. The goal of this book is to show how to solve problems instead of proving theorems. This mischievous and somewhat provocative statement should be understood in the spirit of Peter Lax' comment in an interview given on the occasion of his receiving the 2005 Abel Prize[4]: "When a mathematician says he has solved the problem he means he knows the solution exists, that it's unique, but very often not much more." Going through mathematical proofs is a serious piece of work: we hope that reading this book feels less like work and more like a thought-provoking experience.

The statistical interpretation, and in particular the Bayesian point of view, plays a central role in this book. Why is it so important to emphasize the philosophical difference between statistical and non-statistical approaches to modelling and problem solving? There are two compelling reasons.

The first one is very practical: admitting the lack of information by modelling the unknown parameters as random variables and encoding the nature of uncertainty into probability densities gives a great freedom to develop the models without having to worry too much about whether solutions exist or are unique. The solution in Bayesian statistics, in fact, is not a single value of the unknowns, but a probability distribution of possible values, that always exists. Moreover, there are often pieces of qualitative information available that simply do not yield to classical methods, but which have a natural interpretation in the Bayesian framework.

It is often claimed, in particular by mathematician in inverse problems working with classical regularization methods, that the Bayesian approach is yet another way of introducing regularization into problems where the data are insufficient or of low quality, and that every prior can be replaced by an appropriately chosen penalty. Such statement may seem correct in particular cases when limited computational resources and lack of time force one to use the Bayesian techniques for finding a single value, typically the maximum a posteriori estimate, but in general the claim is wrong. The Bayesian framework, as we shall reiterate over and again in this book, can be used to produce

---

[4] M. Raussen and C. Skau: *Interview with Peter D. Lax.* Notices of the AMS **53** (2006) 223–229.

particular estimators that coincide with classical regularized solutions, but the framework itself does not reduce to these solutions, and claiming so would be an abuse of syllogism[5].

The second, more compelling reason for advocating the Bayesian approach, has to do with the interpretation of mathematical models. It is well understood, and generally accepted, that a computational model is always a simplification. As George E. P. Box noted, "all models are wrong, some are useful". As computational capabilities have grown, an urge to enrich existing models with new details has emerged. This is particularly true in areas like computational systems biology, where the new paradigm is to study the joint effect of a huge number of details[6] rather than using the reductionist approach and seeking simplified lumped models whose behavior would be well understood. As a consequence, the computational models contain so many model parameters that hoping to determine them based on few observations is simply unreasonable. In the old paradigm, one could say that there are some values of the model parameters that correspond in an "optimal way" to what can be observed. The identifiability of a model by idealized data is a classic topic of research in applied mathematics. From the old paradigm, we have also inherited the faith in the power of single outputs. Given a simplistic electrophysiological model of the heart, a physician would want to see the simulated electrocardiogram. If the model was simple, for example two rotating dipoles, that output would be about all the model could produce, and no big surprises were to be expected. Likewise, given a model for millions of neurons, the physician would want to see a simulated cerebral response to a stimulus. But here is the big difference: the complex model, unlike the simple dipole model, can produce a continuum of outputs corresponding to fictitious data, never measured by anybody in the past or the future. The validity of the model is assessed according to whether the simulated output corresponds to what the physician *expects*. While when modelling the heart by a few dipoles, a single simulated output could still make sense, in the second case the situation is much more complicated. Since the model is overparametrized, the system cannot be identified by available or even hypothetical data and it is possible to obtain completely different outputs simply by adjusting the parameters. This observation can lead researchers to state, in frustration, "well, you can make your model do whatever you want, so what's the point"[7]. This sense of hopelessness is exactly what the Bayesian approach seeks to remove. Suppose that the values of the parameters in the

---

[5] A classic example of analogous abuse of logic can be found in elementary books of logic: while it is true that Aristotle is a Greek, it is not true that a Greek is Aristotle.

[6] This principle is often referred to as *emergence*, as new unforeseen and qualitatively different features emerge as a sum of its parts (cf. physics $\longrightarrow$ chemistry $\longrightarrow$ life $\longrightarrow$ intelligence). Needless to say, this holistic principle is old, and can be traced back to ancient philosophers.

[7] We have actually heard this type of statement repeatedly from people who refuse to consider the Bayesian approach to problem solving.

complex model have been set so that the simulated output is completely in conflict with what the experts expect. The reaction to such output would be to think that the current settings of the parameters "must" be wrong, and there would usually be unanimous consensus about the incorrectness of the model prediction. This situation clearly demonstrates that some combinations of the parameter values have to be excluded, and the exclusion principle is based on the observed data (likelihood), or in lack thereof, on the subjective belief of an expert (prior). Thanks to this exclusion, the model can no longer do whatever we want, yet we have not reduced its complexity and thereby its capacity to capture complex, unforeseen, but possible, phenomena. By following the principle of exclusion and subjective learned opinions, we effectively narrow down the probability distributions of the model parameters so that the model produced plausible results. This process is cumulative: when new information arrives, old information is not rejected, as is often the case in the infamous "model fitting by parameter tweaking", but included as prior information. This mode of building models is not only Bayesian, but also Popperian[8] in the wide sense: data is used to *falsify* hypotheses thus leading to the removal of impossible events or to assigning them as unlikely, rather than to verify hypotheses, which is in itself a dubious project. As the classic philosophic argument goes, producing one white swan, or, for that matter, three, does not prove the theory that all swans are white. Unfortunately, deterministic models are often used in this way: one, or three, successful reconstructions are shown as proof of a concept.

The statistical nature of parameters in complex models serves also another purpose. When writing a complex model for the brain, for instance, we expect that the model is, at least to some extent, generic and representative, and thus capable of explaining not one but a whole population of brains. To our grace, or disgrace, not all brains are equal. Therefore, even without a reference to the subjective nature of information, a statistical model simply admits the diversity of those obscure objects of our modelling desires.

This books, which is based on notes for courses that we taught at Case Western Reserve University, Helsinki University of Technology, and at the University of Udine, is a tutorial rather than an in-depth treatise in Bayesian statistics, scientific computing and inverse problems. When compiling the bibliography, we faced the difficult decision of what to include and what to leave out. Being at the crossroad of three mature branches of research, statistics, numerical analysis and inverse problems, we were faced with three vast horizons, as there were three times as many people whose contributions should have been acknowledged. Since compiling a comprehensive bibliography seemed a herculean task, in the end Occam's razor won and we opted to list only the books that were suggested to our brave students, whom we thank for feedback and comments. We also want to thank Dario Fasino for his great hos-

---

[8] See A. Tarantola: *Inverse problems, Popper and Bayes*, Nature Physics **2** 492–494, (2006).

pitality during our visit to Udine, Rebecca Calvetti, Rachael Hageman and Rossana Occhipinti for help with proofreading. The financial support of the Finnish Cultural Foundation for Erkki Somersalo during the completion of the manuscript is gratefully acknowledged.

Cleveland – Helsinki,   *Daniela Calvetti*
May 2007   *Erkki Somersalo*

Contents

1   **Inverse problems and subjective computing** 1
   1.1 What do we talk about when we talk about random variables? 2
   1.2 Through the formal theory, lightly 5
   1.3 How normal is it to be normal? 16

2   **Basic problem of statistical inference** 21
   2.1 On averaging 22
   2.2 Maximum Likelihood, as frequentists like it 31

3   **The praise of ignorance: randomness as lack of information**   39
   3.1 Construction of Likelihood 41
   3.2 Enter, Subject: Construction of Priors 48
   3.3 Posterior Densities as Solutions of Statistical Inverse Problems   55

4   **Basic problem in numerical linear algebra** 61
   4.1 What is a solution? 61
   4.2 Direct linear system solvers 63
   4.3 Iterative linear system solvers 67
   4.4 Ill-conditioning and errors in the data 77

5   **Sampling: first encounter** 91
   5.1 Sampling from Gaussian distributions 92
   5.2 Random draws from non-Gaussian densities 99
   5.3 Rejection sampling: prelude to Metropolis-Hastings 102

6   **Statistically inspired preconditioners** 107
   6.1 Priorconditioners: specially chosen preconditioners 108
   6.2 Sample-based preconditioners and PCA model reduction 118

7   Conditional Gaussian densities and predictive envelopes ... 127
   7.1   Gaussian conditional densities............................128
   7.2   Interpolation, splines and conditional densities 134
   7.3   Envelopes, white swans and dark matter 144

8   More applications of the Gaussian conditioning............147
   8.1   Linear inverse problems 147
   8.2   Aristotelian boundary conditions 151

9   Sampling: the real thing 161
   9.1   Metropolis–Hastings algorithm 168

**10 Wrapping up: hypermodels, dynamic priorconditioners
    and Bayesian learning** 183
   10.1 MAP estimation or marginalization? 189
   10.2 Bayesian hypermodels and priorconditioners 193

References..197

Index...199

1

Inverse problems and subjective computing

The rather provocative subtitle of this book, *Ten Lectures on Subjective Computing*, is intended to evoke the concoction of two relatively well established concepts, *subjective probability* and *scientific computing*. The former is concerned usually with Bayesian statistics; the latter with the design and application of numerical techniques to the solution of problems in science and engineering. This book provides an oblique view of numerical analysis and statistics, with the emphasis on a class of computational problems usually referred to as *inverse problems*.

The need for scientific computing to solve inverse problems has been recognized for a long time. The need for subjective probability, surprisingly, has been recognized even longer. One of the founding fathers[1] of the Bayesian approach to inverse problems, in addition to reverend Bayes[2] himself, is Laplace[3], who used Bayesian techniques, among other things, to estimate the mass of planets from astronomical observations.

To understand the connection between inverse problems and statistical inference, consider the following text book definitions:

- INVERSE PROBLEM: The problem of retrieving information of unknown quantities by indirect observations.

---

[1] In those times potential founding mothers were at home raising children and cooking meals.

[2] Thomas Bayes (1702–1761), in his remarkable, posthumously published , *An Essay towards solving a Problem in the Doctrine of Chances*, Phil. Tras. Royal Soc. London **53**, 370–418 (1763), first analyzed the question of inverse probability, that is, how to determine the probability of an event from observation of outcomes. This has now become a core question of modern science.

[3] Pierre-Simon Laplace (1749–1827): French mathematician and natural scientist, one of the developers of modern probability and statistics, among other achievements. His *Mémoire sur la probabilité des causes par les évènemens* from 1774 shows the Bayes formula in action as we use it today. The English translation as well as a discussion of the Memoir can be found in Stigler, S.M.: Laplace's 1774 Memoir on Inverse Probability, Statistical Science, **1**, 359–363 (1986).

- STATISTICAL INFERENCE: The problem of inferring properties of an unknown distribution from data generated from that distribution.

At first sight, the scopes seem to be rather different. The key words, however, are "information", "unknown" and "distribution". If a quantity is unknown, the information about its value is incomplete. This is an ideal setting for introducing *random variables*, whose distribution tells precisely what information is available and, by complementarity, what is lacking. This definition of randomness is at the very core of this book[4].

Statistics provides a flexible and versatile framework to study inverse problems, especially from the computational point of view, as we shall see. Numerical methods have long been used in statistics, in particular in what is known as *computational statistics*. One of the goals of this book is to enable and nurture a flow in the opposite direction. Paraphrasing the famous appeal of J.F.Kennedy, "Ask not what numerical analysis can do for statistics; ask what statistics can do for numerical analysis". Hence, this book addresses a mixture of topics that are usually taught in numerical analysis or statistical inference courses, with the common thread of looking at them from the point of view of retrieving information from indirect observations by effective numerical methods.

1.1 What do we talk about when we talk about random variables?

The concepts of *random events* and *probabilities* assigned to them have an intuitively clear meaning[5] and yet have been the subject of extensive philosophical and technical discussions. Although it is not the purpose of this book to plunge into this subject matter, it is useful to review some of the central concepts and explain the point of view that we follow.

A random event is the complement of a *deterministic event* in the sense that if a deterministic event is one whose outcome is completely predictable, by complementarity the outcome of a random event is not fully predictable. Hence here *randomness* means *lack of information*. The degree of information, or more generally, our *belief* about the outcome of a random event, is expressed in terms of probability.

That said, it is clear that the concept of probability has a subjective component, since it is a subjective matter to decide what is reasonable to believe, and what previous experience the belief is based on. The concept of probability advocated in this book is called subjective probability[6], which is often

---

[4] A harbinger view to inverse problems and statistical inference as a unified field can be found in the significant article by Tarantola, A. and Valette, B.: *Inverse problems = quest for information*, J. Geophys. **50**, 159–170 (1982).

[5] As Laplace put it, probability is just "common sense reduced to calculations".

[6] An inspirational reference of this topic, see [Je04].

a synonym for *Bayesian probability*. To better understand this concept, and the customary objections against it, let us consider some simple examples.

EXAMPLE 1.1: Consider the random event of tossing a coin. It is quite natural to assume that the *odds*, or relative probabilities of complementary events, of getting heads or tails are equal. This is formally written as

$$P(\text{heads}) = P(\text{tails}) = \frac{1}{2},$$

where P stands for the probability. Such assumption is so common to be taken almost as truth, with no judgemental component in it, similarly as it was once dealt with the flatness of the Earth. If, however, somebody were to disagree, and decided to assign probability 0.4, say, to heads and 0.6 to tails, a justification would certainly be required. Since this is meant as a trivial example of *testing a scientific theory*, there must be a way to falsify the theory[7]. The obvious way to proceed is to test it empirically: a coin is tossed a huge number of times, generating a large experimental sample. Based on this sample, we can assign *experimental probabilities* to heads and tails. The outcome of this experiment will favor one theory over another.

The previous example is in line with the *frequentist* definition of probability: the probability of an event is its relative frequency of occurrence in an asymptotically infinite series of repeated experiments. The statistics based on this view is often referred to as *frequentist statistics*[8], and the most common justification is its objective and empirical nature. But is this really enough of a validation?

EXAMPLE 1.2: The concept of randomness as lack of information is useful in situations where the events in question are non-repeatable. A good example is betting on soccer. A bookmaker assigns odds to the outcomes of a football game of Palermo versus Inter[9], and the amount that a bet wins depends on the odds. Since Palermo and Inter have played each other several times in the past, the new game is hardly an independent realization of the same random event. The bookmaker tries to set the odds so that, no matter what the outcome is, his net loss is non-positive. The

---

[7] This requirement for a scientific theory goes back to Karl Popper (1902–1994), and it has been criticized for not reflecting the way scientific theories work in practice. Indeed, a scientific theory is usually accepted as a probable explanation, if it successfully resists attempts to be falsified, as the theory of flat earth or geocentric universe did for centuries. Thus, scientific truth is really a collective subjective matter.

[8] Frequentist statistics is also called Fisherian statistics, after the English statistician and evolutionary biologist Ronald Fisher (1890–1962).

[9] or, if you prefer, an ice hockey game between Canada and Finland, or a baseball game between the Indians and the Yankees.

subjective experience of the bookmaker is clearly essential for the success of his business.

This example is related to the concept of *coherence* and the definition of probability as worked out by De Finetti[10] In its simplest form, assume that you have a random event with two possible outcomes, "A" and "B", and that you are betting on them against your opponent according to the following rules. If you correctly guess the outcome, you win one dollar; otherwise you pay one dollar to your opponent. Based on your information, you assign the prices, positive but less than one dollar, for betting on each of the outcomes. After your have decided the prices, *your opponent decides* for which outcome you have to bet. The right price for your bets should therefore be such that whichever way the result turns out, you feel comfortable with it.

Obviously, if the frequentist's information about the outcome of earlier events is available, it can be used, and it is likely to help in setting the odds, even if the sample is small. However, the probabilities are well defined even if such sample does not exist. Setting the odds can also be based on computational modelling.

EXAMPLE 1.3: Consider a variant of the coin tossing problem: a thumbtack is tossed, and we want its probability of landing with the spike up. Assuming that no tacks are available to generate an experimental sample, we may try to model the experiment. The convex hull of the tack is a cone, and we may assume that the odds of having the spike up is the ratio of the bottom area to the surface area of the conical surface. Hence, we assign a probability based on a deterministic model. Notice, however, that whether the reasoning we used is correct or not is a judgemental issue. In fact, it is possible also to argue that the odds for spike up or spike down are equal, or we may weigh the odds based on the uneven mass distribution. The position of the spike at landing depends on several factors, some of which may be too complicated to model.

In the last example, the ambiguity is due to the lack of a model that would explain the outcome of the random event. The use of computers has made it possible to *simulate* frequentist's data, leading to a new way of setting the odds for events.

EXAMPLE 1.4: Consider the growth of bacteria in a petri dish, and assume that we have to set the probability for the average bacteria density to be above a certain given level. What we may do is to set up a growth model, partitioning the petri dish area in tiny squares and writing a stochastic model, a "Game of Life[11]", for instance along the following lines. Start

---

[10] Bruno De Finetti (1906–1985): Italian mathematician and statistician, one of the developers of the subjective probability theory. De Finetti's writings on the topic are sometimes rather provocative.

[11] This term refers to the cellular automaton (1970) of John Conway, a British mathematician.

with assigning a probability for the culture to spread from an occupied square to an empty square. If an occupied square is surrounded on all sides by occupied squares, there is a certain probability that competition will kill the culture in that square in the next time step. Once the model has been set up, we may run a simulation starting from different initial configurations, thus creating a fictitious simulated sample on which to base our bet. Again, since the model is just a model, and the intrinsic probabilities of the model are based on our belief, the subjective component is evident. If in doubt on the validity of the model, we may add a random error component to account for our lack of belief in the model. In this way, we introduce a hierarchy of randomness, which turns out to be a useful methodological approach.

Let's conclude this discussion by introducing the notion of *objective chance*. There are situations where it is natural to think that the occurrence of certain phenomena has a probability that is not judgemental. The first example in this chapter could be one of them, as is the tossing of a fair die, where "fair" is defined by means of a circular argument: if a die gives different odds to different outcomes, the die is not fair. The current interpretation of quantum mechanics contains an implicit assumption of the existence of objective chance: the square of the absolute value of the solution of the wave equation defines a probability density, and so the theory predicts the chance. Quantum mechanics, however, is not void of the subjective component via either the interpretation or the interference of the observing subject.

1.2 Through the formal theory, lightly

Regardless of the interpretation, the theory of probability and randomness is widely accepted and presented in a standard form. The formalization of the theory is largely due to Kolmogorov[12]. Since our emphasis is on something besides theoretical considerations, we will present a light version of the fundamentals, spiced with examples and intuitive reasoning. For a rigorous discussion, we shall refer to any standard text on probability theory.

Let's begin with some definitions of quantities and concepts which will be used extensively later. We will not worry too much here about the conditions necessary for the quantities introduced to be well defined. If we write an expression, such as an integral, we will assume that it exists, unless otherwise stated. Questions concerning, e.g., measurability or completeness are likewise neglected.

---

[12] Andrej Nikolaevich Kolmogorov (1903–1987): a Russian mathematician, often refereed to as one of the most prominent figures in the 20th century mathematics and the father of modern probability theory – with emphasis on the word "theory".

Define Ω to be a *probability space* equipped with a *probability measure* P that measures the probability of *events* $E \subset \Omega$. In other words, Ω contains all possible events in the form of its own subsets.

We require that

$$0 \leq P(E) \leq 1,$$

meaning that we express the probability of an event as a number between 0 and 1. An event with probability 1 is the equivalent of a *done deal*[13]; one with probability 0 the exact opposite.

The probability is additive in the sense that if $A \cap B = \emptyset$, $A, B \subset \Omega$,

$$P(A \cup B) = P(A) + P(B).$$

Since Ω contains all events,

$$P(\Omega) = 1, \quad (\text{``something happens''})$$

and

$$P(\emptyset) = 0. \quad (\text{``nothing happens''})$$

Two events, A and B, are *independent*, if

$$P(A \cap B) = P(A)P(B).$$

In this case, the probability of event A *and* of event B is the product of the probability of A times the probability of B. Since a probability is a number smaller than or equal to one, the probability of A and B can never be larger than the probability of either A or B. For example, the probability of getting a head when tossing a fair coin is $\frac{1}{2}$. If we toss the coin twice, the probability of getting two heads is $\frac{1}{2} \cdot \frac{1}{2} = \frac{1}{4}$.

The *conditional probability* of A on B is the probability that A happens *provided* that B happens,

$$P(A \mid B) = \frac{P(A \cap B)}{P(B)}.$$

The probability of A conditioned on B is never smaller than the probability of A and B, since $P(B) \leq 1$, and it can be much larger if $P(B)$ is very small. That corresponds to the intuitive idea that the conditional probability is the probability on the probability subspace where the event B has already occurred. It follows from the definition of independent events that, if A and B are mutually independent,

$$P(A \mid B) = P(A), \quad P(B \mid A) = P(B).$$

The probability space is an abstract entity that is a useful basis for theoretical considerations. In practice, it is seldom constructed explicitly, and it is usually circumvented by the consideration of *probability densities* over the state space.

---

[13] meaning something which certainly happens.

EXAMPLE 1.5: Consider the problem of tossing a die. The state space, i.e., the set of all possible outcomes, is $S = \{1, 2, 3, 4, 5, 6\}$. In this case, we may define the probability space as coinciding with the state space, $\Omega = S$. A particular die is then modeled as a mapping

$$X : \Omega \to S, \quad i \mapsto i, \quad 1 \le i \le 6.$$

and the probabilities of random events, using this *particular die*, are

$$P\{X(i) = i\} = w_i, \text{ with } \sum_{i=1}^{6} w_i = 1.$$

Here and in the sequel, we use the curly brackets with the probability when the event, i.e., the subspace of Ω, is given in the form of a descriptive condition such as $X(i) = i$. The weights w_j define a probability density over the state space.

The above mapping X is a random variable. Since in this book we are mostly concerned about phenomena taking on real values, or more generally, values in $\mathbb{R}^n$, we assume that the state space is $\mathbb{R}$. Therefore we restrict our discussion here to probability densities that are absolutely continuous with respect to the Lebesgue measure over the reals[14].

Given a probability space Ω, a real valued *random variable* X is a mapping

$$X : \Omega \to \mathbb{R}$$

which assigns to each element ω of Ω a real value $X(\omega)$. We call $x = X(\omega)$, $\omega \in \Omega$, a *realization* of X. In summary, a random variable is a function, while its realizations are real numbers.

For each $B \subset \mathbb{R}$,

$$\mu_X(B) = P(X^{-1}(B)) = P\{X(\omega) \in B\}.$$

is called the *probability distribution* of X, i.e., $\mu_X(B)$ is the probability of the *event* $x \in B$. The probability distribution $\mu_X(B)$ measures the size of the subspace of Ω mapped onto B by the random variable X. The probability distribution is linked to the *probability density* π_X via the following integration process:

$$\mu_X(B) = \int_B \pi_X(x) dx. \tag{1.1}$$

From now on, unless it is necessary to specify the dependency on the random variable, we write simply

$$\pi_X(x) = \pi(x).$$

---

[14] Concepts like absolute continuity of measures will play no role in this book, and this comment has been added here in anticipation of "friendly fire" from our fellow mathematicians.

Analogously to what we did in Example 1.5, we identify the probability space Ω and the state space $\mathbb{R}$. It is tempting to identify the probability measure P with the Lebesgue measure. The problem with this is that the measure of $\mathbb{R}$ is infinite, thus the Lebesgue measure cannot represent a probability measure. It is possible – but rather useless for our goals – to introduce an integrable weight on $\mathbb{R}$, defining a probability measure over $\mathbb{R}$, and to integrate with respect to this weight. Since such complication would bring mostly confusion, we stick to the intuitive picture of the Lebesgue measure as the flat reference probability distribution.

Given a random variable X, its *expectation* is the center of mass of the probability distribution,

$$\mathrm{E}\{X\} = \int_{\mathbb{R}} x\pi(x)dx = \overline{x},$$

while its *variance* is the expectation of the squared deviation from the expectation,

$$\mathrm{var}(X) = \sigma_X^2 = \mathrm{E}\{(X - \overline{x})^2\} = \int_{\mathbb{R}} (x - \overline{x})^2\pi(x)dx.$$

It is the role of the variance to measure how distant from the expectation the values taken on by the random variable are. A small variance means that the random variable takes on mostly values close to the expectation, while a large variance means the opposite.

The expectation and variance are the first two moments of a probability density function. The kth moment is defined as

$$\mathrm{E}\{(X - \overline{x})^k\} = \int_{\mathbb{R}} (x - \overline{x})^k\pi(x)dx.$$

The third and fourth moments are closely related to the *skewness* and *kurtosis* of the density.

EXAMPLE 1.6: The expectation of a random variable, in spite of its name, is not necessarily the value that we should expect a realization to take on. Whether this is the case or not depends on the distribution of the random variable. Let $\Omega = [-1, 1]$, the probability P be uniform, in the sense that

$$\mathrm{P}(I) = \frac{1}{2}\int_I dx = \frac{1}{2}|I|, \quad I \subset [-1, 1].$$

and consider the two random variables

$$X_1 : [-1, 1] \rightarrow \mathbb{R}, \quad X_1(\omega) = 1 \quad \forall \omega \in \mathbb{R},$$

and

$$X_2 : [-1, 1] \rightarrow \mathbb{R}, \quad X_2(\omega) = \begin{cases} 2, & \omega \geq 0 \\ 0, & \omega < 0 \end{cases}$$

It is immediate to check that

$$E\{X_1\} = E\{X_2\} = 1,$$

and that while all realizations of X_1 take on the expected value, this is never the case for the realizations of X_2. This simple consideration serves as a warning that describing a random variable only via its expectation may be very misleading. For example, although females make up roughly half of the world's population, and males the other half, we do not expect a randomly picked person to have one testicle and one ovary! By the same token, why should we expect by default a random variable to take on its expected value?

The *joint probability density* of two random variables, X and Y, $\pi_{XY}(x, y)$ $= \pi(x, y)$, is

$$P\{X \in A, Y \in B\} = P(X^{-1}(A) \cap Y^{-1}(B)) = \int\int_{A \times B} \pi(x, y) dx dy$$

$$= \text{the probability of the event that } x \in A$$
$$\text{and, } at\ the\ same\ time,\ y \in B.$$

We say that the random variables X and Y are *independent* if

$$\pi(x, y) = \pi(x)\pi(y),$$

in agreement with the definition of independent events.

The *covariance* of X and Y is the *mixed central moment*

$$\text{cov}(X, Y) = E\{(X - \bar{x})(Y - \bar{y})\}.$$

It is straightforward to verify that

$$\text{cov}(X, Y) = E\{XY\} - E\{X\}E\{Y\}.$$

The *correlation coefficient* of X and Y is

$$\text{corr}(X, Y) = \frac{\text{cov}(X, Y)}{\sigma_X \sigma_Y}, \quad \sigma_X = \sqrt{\text{var}(X)}, \ \sigma_Y = \sqrt{\text{var}(Y)}.$$

or, equivalently, the correlation of the centered normalized random variables,

$$\text{corr}(X, Y) = E\{\tilde{X}\tilde{Y}\}, \quad \tilde{X} = \frac{X - \bar{x}}{\sigma_X}, \quad \tilde{Y} = \frac{Y - \bar{y}}{\sigma_Y}.$$

It is an easy exercise to verify that

$$E\{\tilde{X}\} = E\{\tilde{Y}\} = 0, \quad \text{var}(\tilde{X}) = \text{var}(\tilde{Y}) = 1.$$

The random variables X and Y are *uncorrelated* if their correlation coefficient vanishes, i.e.,

$$\text{cov}(X, Y) = 0.$$

If X and Y are independent variables they are uncorrelated, since

$$\text{E}\{(X - \overline{x})(Y - \overline{y})\} = \text{E}\{X - \overline{x}\}\text{E}\{Y - \overline{y}\} = 0.$$

The viceversa does not necessarily hold. Independency affects the whole density, not just the expectation.

Two random variables X and Y are *orthogonal* if

$$\text{E}\{XY\} = 0.$$

In that case

$$\text{E}\{(X + Y)^2\} = \text{E}\{X^2\} + \text{E}\{Y^2\}.$$

Given two random variables X and Y with joint probability density $\pi(x, y)$, the *marginal density* of X is the probability of X when Y may take on any value,

$$\pi(x) = \int_{\mathbb{R}} \pi(x, y) dy.$$

The marginal of Y is defined analogously by

$$\pi(y) = \int_{\mathbb{R}} \pi(x, y) dx.$$

In other words, the marginal density of X is simply the probability density of X, regardless of Y.

The *conditional probability density* of X given Y is the probability density of X assuming that $Y = y$. By considering conditional probabilities at the limit when Y belongs to an interval shrinking to a single point y, one may show that the conditional density is

$$\pi(x \mid y) = \frac{\pi(x, y)}{\pi(y)}, \quad \pi(y) \neq 0.$$

Observe that, by the symmetry of the roles of X and Y, we have

$$\pi(x, y) = \pi(x \mid y)\pi(y) = \pi(y \mid x)\pi(x), \tag{1.2}$$

leading to the important identity,

$$\pi(x \mid y) = \frac{\pi(y \mid x)\pi(x)}{\pi(y)}, \tag{1.3}$$

known as the *Bayes formula*. This identity will play a central role in the rest of this book.

The interpretations of the marginal and conditional densities are represented graphically in Figure 1.1.

Conditional density leads naturally to *conditional expectation* or *conditional mean*,

$$E\{X \mid y\} = \int_{\mathbb{R}} x\pi(x \mid y)dx.$$

To compute the expectation of X via its conditional expectation, observe that

$$E\{X\} = \int x\pi(x)dx = \int x\left(\int \pi(x,y)dy\right)dx,$$

and, substituting (1.2) into this expression, we obtain

$$E\{X\} = \int x\left(\int \pi(x \mid y)\pi(y)dy\right)dx$$

$$= \int \left(\int x\pi(x \mid y)dx\right)\pi(y)dy = \int E\{X \mid y\}\pi(y)dy. \quad (1.4)$$

So far, we have only considered real valued univariate random variables. A *multivariate random variable* is a mapping

$$X = \begin{bmatrix} X_1 \\ \vdots \\ X_n \end{bmatrix} : \Omega \to \mathbb{R}^n,$$

where each component X_i is a real-valued random variable. The probability density of X is the joint probability density $\pi = \pi_X : \mathbb{R}^n \to \mathbb{R}_+$ of its components, and its expectation is

$$\overline{x} = \int_{\mathbb{R}^n} x\pi(x)dx \in \mathbb{R}^n,$$

or, componentwise,

$$\overline{x}_i = \int_{\mathbb{R}^n} x_i\pi(x)dx \in \mathbb{R}, \quad 1 \le i \le n.$$

The *covariance matrix* is defined as

$$\mathrm{cov}(X) = \int_{\mathbb{R}^n} (x - \overline{x})(x - \overline{x})^{\mathrm{T}}\pi(x)dx \in \mathbb{R}^{n \times n},$$

or, componentwise,

$$\mathrm{cov}(X)_{ij} = \int_{\mathbb{R}^n} (x_i - \overline{x}_i)(x_j - \overline{x}_j)\pi(x)dx \in \mathbb{R}^{n \times n}, \quad 1 \le i, j \le n.$$

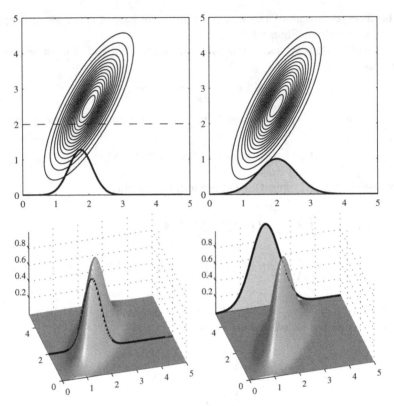

Fig. 1.1. Conditional density (left) and marginal density (right). The ellipsoids in the top panels are equiprobability curves of the joint probability density $\pi(x,y)$. The lower panels show a 3D representation of the densities.

The covariance matrix is *symmetric* and *positive semi-definite*. While the symmetry is implicit in the definition, the positive semi-definiteness follows because, for any $v \in \mathbb{R}^n$, $v \neq 0$,

$$v^{\mathrm{T}}\mathrm{cov}(X)v = \int_{\mathbb{R}^n} \left[v^{\mathrm{T}}(x - \overline{x})\right]\left[(x - \overline{x})^{\mathrm{T}}v\right]\pi(x)dx \qquad (1.5)$$

$$= \int_{\mathbb{R}^n} \left(v^{\mathrm{T}}(x - \overline{x})\right)^2 \pi(x)dx \geq 0.$$

The expression (1.5) measures the variance of X in the direction of v. If the probability density is not *degenerate* in any direction, in the sense that the values of $X - \overline{x}$ are not restricted to any proper subspace, then the covariance is positive definite.

The diagonal entries of the covariance matrix are the variances of the individual components of the random variable X. Denoting by $x'_i \in \mathbb{R}^{n-1}$ the vector x with the ith component deleted, we have

$$\text{cov}(X)_{ii} = \int_{\mathbb{R}^n} (x_i - \bar{x}_i)^2 \pi(x) dx = \int_{\mathbb{R}} (x_i - \bar{x}_i)^2 \underbrace{\left(\int_{\mathbb{R}^{n-1}} \pi(x_i, x_i') dx_i' \right)}_{=\pi(x_i)} dx_i$$

$$= \int_{\mathbb{R}} (x_i - \bar{x}_i)^2 \pi(x_i) dx_i = \text{var}(X_i).$$

The marginal and conditional probabilities for multivariate random variables are defined by the same formulas as in the case of real valued, or univariate, random variables.

We conclude this section with some examples of expectations.

EXAMPLE 1.7: Random variables waiting for the train: assume that every day, except on Sundays, a train for your destination leaves every S minutes from the station. On Sundays, the interval between trains is $2S$ minutes. You arrive at the station with no information about the trains time table. What is your expected waiting time?

Define a random variable, T = waiting time, whose distribution on working days is

$$T \sim \pi(t \mid \text{working day}) = \frac{1}{S} \chi_S(t), \quad \chi_S(t) = \begin{cases} 1, & 0 \leq t < S, \\ 0 & \text{otherwise} \end{cases}$$

On Sundays, the distribution of T is

$$T \sim \pi(t \mid \text{Sunday}) = \frac{1}{2S} \chi_{2S}(t).$$

If you are absolutely sure that it is a working day, the expected waiting time is

$$E\{T \mid \text{working day}\} = \int t \pi(t \mid \text{working day}) dt = \frac{1}{S} \int_0^S t \, dt = \frac{S}{2}.$$

On Sundays, the expected waiting time is clearly double.

If you have no idea which day of the week it is[15], you may give equal probability to each day. Thus,

$$\pi(\text{working day}) = \frac{6}{7}, \quad \pi(\text{Sunday}) = \frac{1}{7}.$$

Since you are interested only in the waiting time for your train, regardless of the day of the week, you marginalize over the days of the week and obtain

$$E\{T\} = E\{T \mid \text{working day}\} \pi(\text{working day}) + E\{T \mid \text{Sunday}\} \pi(\text{Sunday})$$

$$= \frac{3S}{7} + \frac{S}{7} = \frac{4S}{7}.$$

Here we have used formula (1.4), with the integral replaced by a finite sum.

---

[15] The probability of this to happen seems to be much larger for mathematicians than for other individuals.

The following example demonstrates how from simple elementary principles one can construct a probability density that has the characteristics of objective chance, i.e., the resulting probability density follows inarguably from those principles. Naturally, the elementary principles may always be questioned.

EXAMPLE 1.8: Distributions and photon counts. A weak light source emits photons that are counted with a CCD (*Charged Coupled Device*). The *counting process* $N(t)$,

$$N(t) = \text{number of particles observed in } [0, t] \in \mathbb{N}$$

is an integer-valued random variable.

To set up a statistical model, we make the following assumptions:

1. *Stationarity:* Let Δ_1 and Δ_2 be any two time intervals of equal length, and let n be a non-negative integer. Assume that

$$P\{n \text{ photons arrive in } \Delta_1\} = P\{n \text{ photons arrive in } \Delta_2\}.$$

2. *Independent increments:* Let $\Delta_1, \ldots, \Delta_n$ be non-overlapping time intervals and $k_1, \ldots, k_n$ non-negative integers. Denote by A_j the event defined as

$$A_j = k_j \text{ photons arrive in the time interval } \Delta_j.$$

We assume that these events are mutually independent,

$$P\{A_1 \cap \cdots \cap A_n\} = P\{A_1\} \cdots P\{A_n\}.$$

3. *Negligible probability of coincidence:* Assume that the probability of two or more events occurring at the same time is negligible. More precisely, assume that $N(0) = 0$ and

$$\lim_{h \to 0} \frac{P\{N(h) > 1\}}{h} = 0.$$

This condition can be interpreted as saying that the number of counts increases at most linearly.

If these assumptions hold, then it can be shown[16] that N is a *Poisson process:*

$$P\{N(t) = n\} = \frac{(\lambda t)^n}{n!} e^{-\lambda t}, \quad \lambda > 0.$$

We now fix $t = T =$ recording time, define a random variable $N = N(T)$, and let $\theta = \lambda T$. We write

$$N \sim \text{Poisson}(\theta).$$

---

[16] For the proof, see, e.g., [Gh96].

We want to calculate the expectation of this Poisson random variable. Since the discrete probability density is

$$\pi(n) = \mathrm{P}\{N = n\} = \frac{\theta^n}{n!}e^{-\theta}, \quad \theta > 0, \tag{1.6}$$

and our random variable takes on discrete values, in the definition of expectation instead of an integral we have an infinite sum, that is

$$\mathrm{E}\{N\} = \sum_{n=0}^{\infty} n\pi(n) = e^{-\theta}\sum_{n=0}^{\infty} n\frac{\theta^n}{n!}$$

$$= e^{-\theta}\sum_{n=1}^{\infty} \frac{\theta^n}{(n-1)!} = e^{-\theta}\sum_{n=0}^{\infty} \frac{\theta^{n+1}}{n!}$$

$$= \theta e^{-\theta}\underbrace{\sum_{n=0}^{\infty} \frac{\theta^n}{n!}}_{=e^{\theta}} = \theta.$$

We calculate the variance of a Poisson random variable in a similar way, writing first

$$\mathrm{E}\{(N-\theta)^2\} = \mathrm{E}\{N^2\} - 2\theta\underbrace{\mathrm{E}\{N\}}_{=\theta} + \theta^2$$

$$= \mathrm{E}\{N^2\} - \theta^2$$

$$= \sum_{n=0}^{\infty} n^2\pi(n) - \theta^2,$$

and, substituting the expression for $\pi(n)$ from (1.6), we get

$$\mathrm{E}\{(N-\theta)^2\} = e^{-\theta}\sum_{n=0}^{\infty} n^2\frac{\theta^n}{n!} - \theta^2 = e^{-\theta}\sum_{n=1}^{\infty} n\frac{\theta^n}{(n-1)!} - \theta^2$$

$$= e^{-\theta}\sum_{n=0}^{\infty} (n+1)\frac{\theta^{n+1}}{n!} - \theta^2$$

$$= \theta e^{-\theta}\sum_{n=0}^{\infty} n\frac{\theta^n}{n!} + \theta e^{-\theta}\sum_{n=0}^{\infty} \frac{(\theta)^n}{n!} - \theta^2$$

$$= \theta e^{-\theta}\left((\theta + 1)e^{\theta}\right) - \theta^2$$

$$= \theta,$$

that is, the mean and the variance coincide.

1.3 How normal is it to be normal?

A special role in stochastics and statistics is played by *normal* or *normally distributed* random variables. Another name for normal is Gaussian. We start by giving the definition of normal random variable and then proceed to see why normality is so widespread.

A random variable $X \in \mathbb{R}$ is *normally distributed*, or Gaussian,

$$X \sim \mathcal{N}(x_0, \sigma^2),$$

if

$$P\{X \leq t\} = \frac{1}{\sqrt{2\pi\sigma^2}} \int_{-\infty}^{t} \exp\left(-\frac{1}{2\sigma^2}(x - x_0)^2\right) dx.$$

It can be shown that

$$E\{X\} = x_0, \qquad \text{var}(X) = \sigma^2.$$

We now extend this definition to the multivariate case. A random variable $X \in \mathbb{R}^n$ is Gaussian if its probability density is

$$\pi(x) = \left(\frac{1}{(2\pi)^n \det(\Gamma)}\right)^{1/2} \exp\left(-\frac{1}{2}(x - x_0)^{\mathsf{T}} \Gamma^{-1}(x - x_0)\right),$$

where $x_0 \in \mathbb{R}^n$, and $\Gamma \in \mathbb{R}^{n \times n}$ is symmetric positive definite. The expectation of this multivariate Gaussian random variable is x_0 and its covariance matrix is Γ. A *standard normal* random variable is a Gaussian random variable with zero mean and covariance the identity matrix.

Gaussian random variables arise naturally when *macroscopic* measurements are averages of individual *microscopic* random effects. They arise commonly in the modeling of pressure, temperature, electric current and luminosity. The ubiquity of Gaussian random variables can be understood in the light of the Central Limit Theorem, which we are going to state here for completeness, without presenting a proof[17].

Central Limit Theorem: Assume that real valued random variables $X_1, X_2, \ldots$ are *independent* and *identically distributed* (i.i.d.), each with expectation μ and variance σ^2. Then the distribution of

$$Z_n = \frac{1}{\sigma\sqrt{n}}(X_1 + X_2 + \cdots + X_n - n\mu)$$

converges to the distribution of a standard normal random variable,

$$\lim_{n \to \infty} P\{Z_n \leq x\} = \frac{1}{2\pi} \int_{-\infty}^{x} e^{-t^2/2} dt.$$

---

[17] The proof can be found in standard text books, e.g., [Gh96].

Another way of thinking about the result of the Central Limit Theorem is the following. If

$$Y_n = \frac{1}{n} \sum_{j=1}^{n} X_j,$$

and n is large, a good approximation for the probability distribution of Y is

$$Y_n \sim \mathcal{N}\left(\mu, \frac{\sigma^2}{n}\right).$$

As an example of the usefulness of this result, consider a container filled with gas. The moving gas molecules bombard the walls of the container, the net effect being the observed pressure of the gas. Although the velocity distribution of the gas molecules is unknown to us, the net effect, which is the average force per unit area is, by the Central Limit Theorem, Gaussian.

EXAMPLE 1.9: Consider again the CCD camera of the previous example with a relatively large number of photons. We may think of dividing the counter unit into smaller sub-counters, so that the total count is the sum of the sub-counts. Since all sub-counters are identical and the photon source is the same for each of them, it is to be expected that the total count is a sum of identical Poisson processes. Thus it is reasonable to expect that, when the mean θ is large, the Poisson distribution can be approximated well by a Gaussian distribution.

If this reasoning is correct, the approximating Gaussian distribution should have asymptotically the same mean and variance as the underlying Poisson distribution. In Figure 1.2 we plot the Poisson probability density

$$n \mapsto \frac{\theta^n}{n!} e^{-\theta} = \pi_{\text{Poisson}}(n \mid \theta)$$

versus the Gaussian approximation

$$x \mapsto \frac{1}{\sqrt{2\pi\theta}} \exp\left(-\frac{1}{2\theta}(x - \theta)^2\right) = \pi_{\text{Gaussian}}(x \mid \theta, \theta),$$

for increasing values of the mean. The approximation, at least visually, becomes more accurate as the mean increases. Notice, however, that since the Poisson distribution is skewed, matching the means does not mean that the maxima are matched, nor does it give a measure for the goodness of the fit.

To obtain a quantitative measure for the goodness of the approximation, we may consider metric distances between the probability densities. Several metrics for the distance between probability distributions exist. One often used in engineering literature,

$$\text{dist}_{\text{KL}}\left(\pi_{\text{Poisson}}(\,\cdot\,\mid \theta), \pi_{\text{Gaussian}}(\,\cdot\,\mid \theta, \theta)\right)$$

$$= \sum_{n=0}^{\infty} \pi_{\text{Poisson}}(n \mid \theta) \log\left(\frac{\pi_{\text{Poisson}}(n \mid \theta)}{\pi_{\text{Gaussian}}(n \mid \theta, \theta)}\right),$$

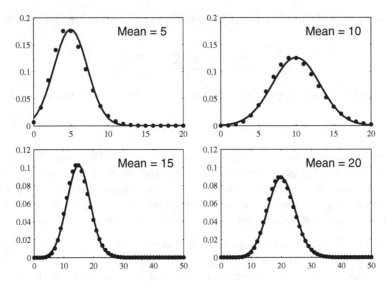

Fig. 1.2. Poisson distributions (dots) and their Gaussian approximations (solid line) for various values of the mean θ.

is known as the *Kullback–Leibler distance.*

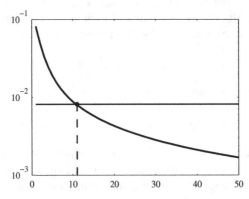

Fig. 1.3. The Kullback-Leibler distance of the Poisson distribution from its Gaussian approximation as a function of the mean θ in logarithmic scale. The horizontal line indicates where the distance has dropped to one tenth of its value at $\theta = 1$.

Figure 1.3 shows the Kullback-Leibler distance as a function of the parameter θ. The often quoted rule of thumb is that when $\theta \geq 15$, it is reasonable to approximate the Poisson distribution with a Gaussian.

The Central Limit Theorem justifies in many circumstances the use of Gaussian approximations. As we shall see, normal distributions are computationally quite convenient. There is also a cognitive dimension which plays an important role in the widespread use of Gaussian approximations, as illustrated by the following example.

EXAMPLE 1.10: An experienced radiologist decides to quantify her qualitative knowledge of typical radiographs of the human thorax. The extensive collection of her memories of such radiographs all display the same typical features: a heart, two lungs, the ribcage. It seems like in her mind the collective memories of the radiographs have melted into an average image[18], with the understanding that there could be slight variations around it. Thus an efficient way to summarize her qualitative experience in quantitative terms is to describe the average image and the degree of variation around it. It is not unreasonable then to write a Gaussian model for a typical thoracic radiograph, as an expression of the mental averaging process. This description is clearly subjective, as it relies on personal experiences.

Exercises

1. Consider the following modification of Example 1.7. Instead of every S minutes, the train leaves from the station every day at the same, unevenly distributed times, $t_j, 1 \leq j \leq n$. Assume that you have the train time table but no watch[19], nor any idea of what time of the day it is. What is the expected waiting time in this case?

2. Consider a vector valued random variable,

$$A = Xe_1 + Ye_2,$$

where e_1 and e_2 are the orthogonal Cartesian unit vectors, and X and Y are real valued random variables,

$$X, Y \sim \mathcal{N}(0, \sigma^2).$$

The random variable
$$R = \|A\|$$

is then distributed according to the *Rayleigh distribution*,

---

[18] The cognitive dimension of averaging visual features has clearly emerged in studies where digitized images of human faces are averaged and then assessed visually by people. It turns out that an average face is typically found attractive, an easily justified outcome from the evolutionary perspective. The result seems to suggest that humans, in some sense, *recognize* the average, and, therefore, the assumption that the brain averages images is not unnatural. See, e.g., Langlois, J.H. and Roggman, L.A.: *Attractive faces are only average*. Psychological Science **1**, 115–121 (1990).

[19] The situation is not as bad as having a loaf of bread with a hard crust and no teeth to eat it with.

$$R \sim \text{Rayleigh}(\sigma^2).$$

Derive the analytic expression of the Rayleigh distribution and write a Matlab program that generates points from the Rayleigh distribution. Make a plot of the distribution and a histogram of the points you generated.

3. Calculate the third and fourth moments of a Gaussian density.

2

Basic problem of statistical inference

The definition of statistical inference given at the beginning of Chapter 1 suggests that the *fundamental problem of statistical inference* can be stated as follows:

Given a set of observed realizations of a random variable X,

$$S = \{x_1, x_2, \ldots, x_N\}, \quad x_j \in \mathbb{R}^n, \tag{2.1}$$

we want to infer on the underlying probability distribution that gives rise to the data S.

An essential portion of statistical inference deals with *statistical modelling* and *analysis*, which often includes the description and understanding of the data collection process. Therefore, statistical inference lies at the very core of scientific modeling and empirical testing of theories.

It is customary to divide the various approaches in two categories;

- *Parametric approach*: It is assumed that the underlying probability distribution giving rise to the data can be described in terms of a number of parameters, thus the inference problem is to estimate these parameters, or possibly their distributions.
- *Non-parametric approach:* No assumptions are made about the specific functional form of the underlying probability density. Rather, the approach consists of describing the dependency or independency structure of the data, usually in conjunction with numerical exploration of the probability distribution.

In classical statistical problems, parametric models depend on relatively few parameters such as the mean and the variance. If the data set is large, this is equivalent to reducing the model, in the sense that we attempt to explain a large set of values with a smaller one. However, when we apply parametric methods to inverse problems, this may not be the case. The parametric model typically depends on the unknowns of primary interest, and the number of degrees of freedom may exceed the size of the data.

Nonparametric models have the advantage of being less sensitive to mis-modelling the genesis of the data set. The disadvantages include the fact that, in the absence of a synthetic formula, it is often hard to understand the factors that affect the data.

The basic problem of statistical inference as stated above is non-committal with respect to the interpretation of probability. It is interesting to notice that, even within the parametric modelling approach, the methodology of statistical inference is widely different, depending on whether one adheres to the frequentist point of view of probability, or to the Bayesian paradigm[1]. In the framework of parametric models, one may summarize the fundamental difference by saying that a frequentist believes in the existence of underlying true parameter values, while in the Bayesian paradigm, anything that is not known, *including the parameters*, must be modeled as a random variable, randomness being an expression of lack of information, or, better said, of *ignorance*[2]. Hence, there is a Platonist aspect in the frequentist's approach, while the Bayesian approach is closer to the Aristotelian concept of learning as starting from a clean slate. As Aristotle learned a lot from his mentor Plato, we can also learn a lot from the frequentist's approach, which will be our starting point.

2.1 On averaging

The most elementary way to summarize a large set of statistical data is to calculate its average. There is a rationale behind this, summarized by the Law of Large Numbers. This law has several variants, one of which is the following.

Law of Large Numbers: Assume that $X_1, X_2, \ldots$ are independent and identically distributed random variables with finite mean μ and variance σ^2. Then

$$\lim_{n \to \infty} \frac{1}{n}(X_1 + X_2 + \cdots + X_n) = \mu$$

almost certainly.

The expression *almost certainly* means that, with probability one, the averages of any realizations $x_1, x_2, \ldots$ of the random variables $X_1, X_2, \ldots$ converge towards the mean. This is good news, since data sets are always realizations. Observe the close connection with the Central Limit Theorem, which states the convergence in terms of probability densities.

---

[1] We recommend the reader to pick up any statistics journal and appreciate the remarkable difference in style of writing, depending on the approach of the author.

[2] We use the word ignorance in this case to indicate lack of knowledge, which goes back to its original Greek meaning.

The following example is meant to clarify how much, or how little, averages over samples are able to explain the underlying distribution.

EXAMPLE 2.1: Given a sample S of vectors $x_j \in \mathbb{R}^2$, $1 \le j \le n$, in the plane, we want to set up a parametric model for it, assuming that the points are independent realizations of a normally distributed random variable,

$$X \sim \mathcal{N}(x_0, \Gamma),$$

with unknown mean $x_0 \in \mathbb{R}^2$ and covariance matrix $\Gamma \in \mathbb{R}^{2\times 2}$. We express this by writing the probability density of X in the form

$$\pi(x \mid x_0, \Gamma) = \frac{1}{2\pi\det(\Gamma)^{1/2}} \exp\left(-\frac{1}{2}(x-x_0)^{\mathrm{T}} \Gamma^{-1}(x-x_0)\right),$$

i.e., the probability density has the known form *provided that x_0 and Γ are given*. The inference problem is therefore to estimate the parameters x_0 and Γ.

As the Law of Large Numbers suggests, we may hope that n is sufficiently large to justify an approximation of the form

$$x_0 = \mathrm{E}\{X\} \approx \frac{1}{n} \sum_{j=1}^n x_j = \widehat{x}_0. \tag{2.2}$$

To estimate the covariance matrix, observe first that if $X_1, X_2, \ldots$ are independent and identically distributed, so are $f(X_1), f(X_2), \ldots$ for any function $f : \mathbb{R}^2 \mapsto \mathbb{R}^k$, therefore we may apply the Law of Large Numbers again. In particular, we may write

$$\Gamma = \mathrm{cov}(X) = \mathrm{E}\{(X-x_0)(X-x_0)^{\mathrm{T}}\}$$
$$\approx \mathrm{E}\{(X-\widehat{x}_0)(X-\widehat{x}_0)^{\mathrm{T}}\} \tag{2.3}$$
$$\approx \frac{1}{n} \sum_{j=1}^n (x_j - \widehat{x}_0)(x_j - \widehat{x}_0)^{\mathrm{T}} = \widehat{\Gamma}.$$

The estimates $\hat{x}$ and $\hat{\Gamma}$ from formulas (2.2) and (2.3) are often referred to as the *empirical mean and covariance*, respectively.

In the first numerical simulation, let us assume that the Gaussian parametric model can indeed explain the sample, which has in fact been generated by drawing from a two dimensional Gaussian density.

The left panel of Figure 2.1 shows the equiprobability curves of the original Gaussian probability density, which are ellipses. A scatter plot of a sample of 200 points drawn from this density is shown in the right panel of the same figure. We then compute the eigenvalue decomposition of the empirical covariance matrix,

$$\widetilde{\Gamma} = UDU^{\mathrm{T}}, \tag{2.4}$$

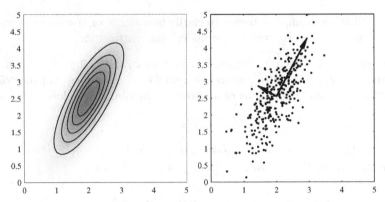

Fig. 2.1. The ellipses of the equiprobability curves of the original Gaussian distribution (left), and on the right, the scatter plot of the random sample and the eigenvectors of the empirical covariance matrix, scaled by two times the square root of the corresponding eigenvalues. These square roots represent the standard deviations of the density in the eigendirections.

where $U \in \mathbb{R}^{2 \times 2}$ is an orthogonal matrix whose columns are the eigenvectors scaled to have unit length, and $D \in \mathbb{R}^{2 \times 2}$ is the diagonal matrix of the eigenvalues of $\tilde{\Gamma}$,

$$ U = \begin{bmatrix} v_1 & v_2 \end{bmatrix}, \quad D = \begin{bmatrix} \lambda_1 & \\ & \lambda_2 \end{bmatrix}, \quad \tilde{\Gamma} v_j = \lambda_j v_j \quad j = 1, 2. $$

The right panel of Figure 2.1 shows the eigenvectors of $\tilde{\Gamma}$ scaled to have length twice the square root of the corresponding eigenvalues, applied at the empirical mean. The plot is in good agreement with our expectation. To better understand the implications of adopting a parametric form for the underlying density, we now consider a sample from a density that is far from Gaussian. The equiprobability curves of this non-Gaussian density are shown in the left panel of Figure 2.2. The right panel shows the scatter plot of a sample of 200 points drawn from this density. We calculate the empirical mean, the empirical covariance matrix and its eigenvalue decomposition. As in the Gaussian case, we plot the eigenvectors scaled by a factor twice the square root of the corresponding eigenvalues, applied at the empirical mean. While the mean and the covariance estimated from the sample are quite reasonable, a Gaussian model is clearly not in agreement with the sample. Indeed, if we would generate a new sample from a Gaussian density with the calculated mean and covariance, the scatter plot would be very different from the scatter plot of the sample.

In two dimensions, it is relatively easy to check whether a Gaussian approximation is reasonable by visually inspecting scatter plots. Although in higher dimensions, we can look at the scatter plots of two-dimensional

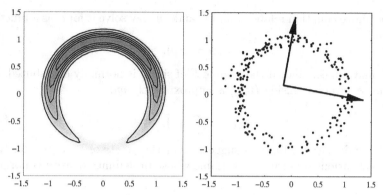

Fig. 2.2. The equiprobability curves of the original non-Gaussian distribution (left) and the sample drawn from it (right), together with the scaled eigenvectors of the empirical covariance matrix applied at the empirical mean.

projections and try to assess if the normality assumption is reasonable[3], it is nonetheless preferable to develop systematic methods to investigate the goodness of the Gaussian approximation. Below, we present one possible idea.

Let $\pi(x)$ denote the Gaussian probability density,

$$\pi \sim \mathcal{N}(\widehat{x}_0, \widehat{\Gamma}),$$

where the mean and the covariance are obtained from a given sample. Consider sets of the form

$$B_\alpha = \{x \in \mathbb{R}^2 \mid \pi(x) \geq \alpha\}, \quad \alpha > 0.$$

Since π is a Gaussian density, the set B_α is either empty (α large) or the interior of an ellipse. Assuming that π is the probability density of the random variable X, we can calculate the probability that X is in B_α by evaluating the integral

$$P\{X \in B_\alpha\} = \int_{B_\alpha} \pi(x)dx. \tag{2.5}$$

We call B_α the *credibility ellipse*[4] with credibility p, $0 < p < 1$, if

$$P\{X \in B_\alpha\} = p. \tag{2.6}$$

---

[3] As the legendary Yogi Berra summarized in one of his famous aphorisms, "you can observe a lot of things just by looking at them".

[4] We choose here to use the Bayesian concept of *credibility* instead of the frequentist concept of *confidence*, since it is more in line with the point of view of this book. Readers confident enough are welcome to choose differently.

The above equation relates α and p, and we may solve it for α as a function of p,

$$\alpha = \alpha(p).$$

We may expect that, if the sample S of size n is normally distributed, the number of points inside $B_{\alpha(p)}$ is approximately pn,

$$\#\{x_j \in B_{\alpha(p)}\} \approx pn. \tag{2.7}$$

A large deviation from this suggests that the points of S are not likely to be realizations of a random variable whose probability density is normal.

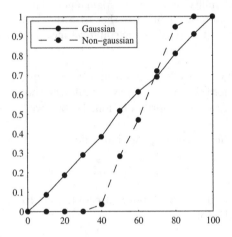

Fig. 2.3. Percentage of sample points inside the credibility ellipse of the approximating Gaussian distribution plotted against the sample size.

To calculate the number of sample points within $B_{\widehat{\alpha}}$, we start by noticing that, replacing $\widehat{\Gamma}$ with its eigenvalue decomposition (2.4),

$$(x - \widehat{x}_0)^{\mathrm{T}} \widehat{\Gamma}^{-1} (x - \widehat{x}_0) = (x - \widehat{x}_0)^{\mathrm{T}} U D^{-1} U^{\mathrm{T}} (x - \widehat{x}_0)$$
$$= \|D^{-1/2} U^{\mathrm{T}} (x - \widehat{x}_0)\|^2,$$

where

$$D^{-1/2} = \begin{bmatrix} 1/\sqrt{\lambda_1} & \\ & 1/\sqrt{\lambda_2} \end{bmatrix}.$$

After the change of variable

$$w = f(x) = W(x - \widehat{x}_0), \quad W = D^{-1/2} U^{\mathrm{T}},$$

the integral (2.5) becomes

$$\int_{B_\alpha} \pi(x) dx = \frac{1}{2\pi \left(\det(\widehat{\Gamma})\right)^{1/2}} \int_{B_\alpha} \exp\left(-\frac{1}{2}(x - \widehat{x}_0)^T \widehat{\Gamma}^{-1}(x - \widehat{x}_0)\right) dx$$

$$= \frac{1}{2\pi \left(\det(\widehat{\Gamma})\right)^{1/2}} \int_{B_\alpha} \exp\left(-\frac{1}{2}\|W(x - \widehat{x}_0)\|^2\right) dx$$

$$= \frac{1}{2\pi} \int_{f(B_\alpha)} \exp\left(-\frac{1}{2}\|w\|^2\right) dw,$$

since

$$dw = \det(W) dx = \frac{1}{\sqrt{\lambda_1 \lambda_2}} dx = \frac{1}{\det(\widehat{\Gamma})^{1/2}} dx.$$

The equiprobability curves for the density of w are circles centered at the origin,

$$f(B_\alpha) = D_\delta = \{w \in \mathbb{R}^2 \mid \|w\| < \delta\}$$

for some $\delta > 0$ that depends on α. The value of δ can be determined directly from (2.6). In fact, passing to radial coordinates (r, θ), we have

$$\frac{1}{2\pi} \int_{D_\delta} \exp\left(-\frac{1}{2}\|w\|^2\right) dw = \int_0^\delta \exp\left(-\frac{1}{2}r^2\right) r dr$$

$$= 1 - \exp\left(-\frac{1}{2}\delta^2\right) = p,$$

which implies that

$$\delta = \delta(p) = \sqrt{2 \log\left(\frac{1}{1 - p}\right)}.$$

To see if a sample point x_j lies inside the p-credibility ellipse, it suffices to check if

$$\|w_j\| < \delta(p), \quad w_j = W(x_j - \widehat{x}_0), \quad 1 \le j \le n,$$

holds.

Figure 2.3 shows the plots of the function

$$p \mapsto \frac{1}{n}\#\{x_j \in B_{\alpha(p)}\}$$

for two samples, one generated from the Gaussian density of Figure 2.1 and the other from the non-Gaussian one of Figure 2.2. The non-Gaussianity of the latter sample can be deduced from this plot, in particular when the value of p is small.

We conclude with the portion of a Matlab code that was used for the calculations. In the code, S is a $2 \times n$ matrix whose columns are the coordinates of the sample points. The generation of the data is not discussed here, since drawing samples from a given density will be discussed at length in Chapters 5 and 9.

```
n = length(S(1,:));          % Size of the sample
xmean = (1/n)*(sum(S')');     % Mean of the sample
CS = S - xmean*ones(1,n);     % Centered sample
Gamma = 1/n*CS*CS';           % Covariance matrix

% Mapping the the x-sample to w-sampe

[V,D] = eig(Gamma);          % Eigenvalue decomposition
W = diag([1/sqrt(D(1,1));1/sqrt(D(2,2))])*V';
WS = W*CS;                    % w-sample matrix

% Calculating the percentage of scatter points inside
% the credibility ellipses

normWS2 = sum(WS.^2);
rinside = zeros(11,1);
rinside(11) = n;
for j = 1:9
    delta2 = 2*log(1/(1-j/10));
    rinside(j+1) = sum(normWS2<delta2);
end
rinside = (1/n)*rinside;

plot([0:10:100],rinside,'k.-','MarkerSize',12)
```

Notice that in the code the sample covariance is calculated from (2.3) instead of using the mathematically equivalent formula

$$\widehat{\Gamma} = \frac{1}{n}\sum_{j=1}^{n} x_j x_j^{\mathrm{T}} - \widetilde{x}_0 \widetilde{x}_0^{\mathrm{T}}. \qquad (2.8)$$

Although for the present example either formula could have been used, in higher dimensional problems, and in particular if the sample is highly correlated, the computational performance of the two formulas might be quite different. The matrix obtained from (2.3) is more likely to be numerically positive definite, while in some cases the matrix generated using formula (2.8) may be so severely affected by round-off errors to become numerically singular or indefinite.

Averaging of data is a common practice in science and engineering. The motivation for averaging may be to summarize an excessive amount of the data in comprehensible terms, but often also noise reduction[5], which is briefly discussed in the following example.

---

[5] Unfortunately, in many cases the motivation is just to represent the data disguised as statistical quantities, while in reality the original data would be a lot more informative.

EXAMPLE 2.2: Assume that we have a measuring device whose output is a noisy signal. This means that the signal generated contains a noise component. In order to estimate the noise level, we perform *calibration* by attaching the measurement device to a dummy load whose theoretical noiseless output we know. Subtracting the theoretical noise from the actual output of the device we obtain a pure noise signal. Assume that the device samples the continuous output signal, and the sampled signal is a vector of given length n. When we perform the calibration measurement, the output is thus a noise vector. In setting up a parametric model, we assume that the noise vector $x \in \mathbb{R}^n$ is a realization of a random variable,

$$X : \Omega \to \mathbb{R}^n.$$

To estimate the noise level, we assume that the components of X are mutually independent, identically distributed random variables. The average noise and its variance, based on a single measurement vector, can be estimated via the formulas

$$x_0 = \frac{1}{n} \sum_{j=1}^{n} x_j, \quad \sigma^2 = \frac{1}{n} \sum_{j=1}^{n} (x_j - x_0)^2.$$

Assume that, after having performed the calibration measurement, we realize that the signal-to-noise ratio of our device is so poor that important features of the signal are cluttered under the noise. In an effort to reduce the noise level, we repeat the measurement several times and average the noisy signals, in the hope of reducing the variance of the noise component. How many measurements do we need to reach a desired signal-to-noise ratio? If $x^{(k)} \in \mathbb{R}^n$ denotes the noise vector in the kth repeated measurement, the noise vector in the averaged signal is

$$x = \frac{1}{N} \sum_{k=1}^{N} x^{(k)} \in \mathbb{R}^n.$$

Assuming that the repeated measurements are mutually independent and the measurement conditions from one measurement to another do not change, we may still view the vector x as an average of independent identically distributed random variables. Since the Central Limit Theorem asserts that asymptotically, as N goes to infinity, the average is Gaussian and its variance goes to zero like σ^2/N, to obtain a signal whose noise variance is below a threshold value τ^2, we need to repeat the measurements N times, where

$$\frac{\sigma^2}{N} < \tau^2.$$

To demonstrate the noise reduction by averaging, we generate Gaussian noise vectors of length $n = 50$ with variance $\sigma^2 = 1$, and average them. In

Figure 2.4 we show averages of 1, 5 and 25 of such signals. We also indicate the noise level, estimated as

$$\text{noise level} = 2\frac{\sigma}{\sqrt{N}},$$

where N is the number of averaged noise vectors and σ is the standard deviation of the single noise vectors.

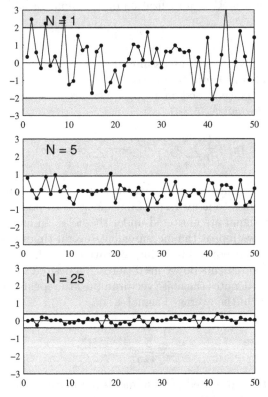

Fig. 2.4. Averaged noise signals. The shading indicates where the noise level is twice the estimated standard deviation, which in this case is $2/\sqrt{N}$.

The mean and variance estimation by averaging relies solely on the formulas suggested by the Law of Large Numbers, and no information concerning the estimated distribution is used. The following section adds more complexity to the problem and discusses maximum likelihood estimators, a central concept in frequentist statistics.

2.2 Maximum Likelihood, as frequentists like it

Assume that we can observe the realizations of a random variable X which depends, in turn, on a parameter θ, which we cannot observe but whose values are really what we are interested in. The parameter can be a real number, a vector, or an array of real numbers. We interpret the sample S as a collection of realizations of a random variable

$$X \sim \pi(x \mid \theta), \quad \theta \in \mathbb{R}^k,$$

where we use the notation of conditional probability density, anticipating the Bayesian framework, where the parameter is also a realization of a random variable,[6] although, for the time being, we take the frequentists' perspective that there is an underlying "true" value of the parameter to be estimated. In frequentist statistics one estimates parameters, while in the Bayesian statistics one infers on their distributions based on one's own beliefs and on available information.

We assume that the observed values x_j of X are independent realizations. This means that the outcome of each single x_j is not affected by the knowledge of the other outcomes. Since it may be confusing to talk about independent realizations of the *same* random variable X, we specify that what we really mean is that we think that there are N independent identically distributed random variables $X_1, X_2, \ldots, X_N$, and that x_j is a realization of X_j. Therefore, referring to a single variable X is just a convenient shorthand notation[7].

The assumption of independence implies that

$$\pi(x_1, x_2, \ldots, x_N \mid \theta) = \pi(x_1 \mid \theta)\pi(x_2 \mid \theta) \cdots \pi(x_N \mid \theta),$$

or, briefly,

$$\pi(S \mid \theta) = \prod_{j=1}^{N} \pi(x_j \mid \theta),$$

where S is the sample (2.1) at our disposal. This can be seen as a measure, in some sense, of the probability of the outcome S, given the value of the parameter θ. The *Maximum Likelihood* (ML) estimator of θ is defined in a very intuitive way as the parameter value that maximizes the probability of the outcomes x_j, $1 \leq j \leq N$:

$$\theta_{\mathrm{ML}} = \arg\max \prod_{j=1}^{N} \pi(x_j \mid \theta),$$

---

[6] The notation more in line with the frequentist tradition would be $\pi_\theta(x)$.

[7] One can think of X as a die that is cast repeatedly, and X_k as the *process* of casting the die the kth time.

provided that such maximizer exists. In other words, without making any assumptions about the parameter to estimate, we take the value that makes the available output *most likely*[8].

We now observe the Maximum Likelihood estimator at work in a couple of applications. We begin with the case where we believe that the data are realizations of independent identically distributed Gaussian random variables, whose expectation and variance we want to estimate.

EXAMPLE 2.3: Consider a real random variable X with a Gaussian distribution

$$\pi(x \mid x_0, \sigma^2) = \frac{1}{\sqrt{2\pi\sigma^2}} \exp\left(\frac{1}{2\sigma^2}(x - x_0)^2\right),$$

whose unknown expectation x_0 and variance σ^2 we want to estimate from a collection $\{x_1, \ldots, x_N\}$ of realizations of X. We begin with collecting the parameters x_0 and σ^2 that we want to estimate into the vector $\theta \in \mathbb{R}^2$,

$$\theta = \begin{bmatrix} x_0 \\ \sigma^2 \end{bmatrix} = \begin{bmatrix} \theta_1 \\ \theta_2 \end{bmatrix},$$

and we define the likelihood function to be

$$\prod_{j=1}^{N} \pi(x_j \mid \theta) = \left(\frac{1}{2\pi\theta_2}\right)^{N/2} \exp\left(-\frac{1}{2\theta_2}\sum_{j=1}^{N}(x_j - \theta_1)^2\right)$$

$$= \exp\left(-\frac{1}{2\theta_2}\sum_{j=1}^{N}(x_j - \theta_1)^2 - \frac{N}{2}\log\left(2\pi\theta_2\right)\right)$$

$$= \exp\left(-L(S \mid \theta)\right).$$

The maximum likelihood estimate of θ is the maximizer of $\exp\left(-L(S \mid \theta)\right)$, or, equivalently, the minimizer of $L(S \mid \theta)$, the negative of the log-likelihood function.

It follows from the differentiability of the map $\theta \mapsto L(S \mid \theta)$ that the minimizer must be a solution of the equation

$$\nabla_\theta L(S \mid \theta) = 0.$$

Since

---

[8] Here, we are edging dangerous waters: it is not uncommon to say that the maximum likelihood estimator is the "most likely" value of the parameter in the light of the data, which is, of course, nonsense *unless* the parameter is also a random variable. This is the key demarcation line between the frequentist and the Bayesian interpretation.

$$\nabla_\theta L(S \mid \theta) = \begin{bmatrix} \dfrac{\partial L}{\partial \theta_1} \\[2ex] \dfrac{\partial L}{\partial \theta_2} \end{bmatrix} = \begin{bmatrix} -\dfrac{1}{\theta_2} \sum_{j=1}^{N} x_j + \dfrac{N}{\theta_2}\theta_1 \\[2ex] -\dfrac{1}{2\theta_2^2} \sum_{j=1}^{N}(x_j - \theta_1)^2 + \dfrac{N}{2\theta_2} \end{bmatrix},$$

solving the first equation for x_0 we have

$$\tilde{x}_0 = \theta_{ML,1} = \frac{1}{N} \sum_{j=1}^{N} x_j,$$

which, substituted in the second equation, gives

$$\tilde{\sigma}^2 = \theta_{ML,2} = \frac{1}{N} \sum_{j=1}^{N}(x_j - \theta_{ML,1})^2.$$

Thus, the Maximum Likelihood estimate of θ_1 is the arithmetic mean of the sample, and the Maximum Likelihood estimate of the variance is the average of the squares of the distances of the data from the estimated expectation. Comparing these findings to the results of the previous subsection we notice that the Maximum Likelihood estimators in the Gaussian case coincide with the experimental mean and variance formulas suggested by the Law of Large Numbers.

Let us now consider an example dealing with a discrete probability density.

EXAMPLE 2.4: Consider a random variable K which has a Poisson distribution with parameter θ,

$$\pi(k) = \frac{\theta^k}{k!} e^{-\theta},$$

and a sample $S = \{k_1, \cdots, k_N\}$, $k_n \in \mathbb{N}$, of independent realizations. The likelihood density is

$$\pi(S \mid \theta) = \prod_{n=1}^{N} \pi(k_n) = e^{-N\theta} \prod_{n=1}^{N} \frac{\theta^{k_n}}{k_n!},$$

and its negative logarithm is

$$L(S \mid \theta) = -\log \pi(S \mid \theta) = \sum_{n=1}^{N} \left(\theta - k_n \log \theta + \log k_n! \right).$$

As in the previous example, we maximize the likelihood by minimizing L. Setting the derivative of the likelihood with respect to θ to zero

$$\frac{\partial}{\partial\theta} L(S \mid \theta) = \sum_{n=1}^{N} \left(1 - \frac{k_n}{\theta}\right) = 0, \tag{2.9}$$

yields

$$\theta_{\text{ML}} = \frac{1}{N} \sum_{n=1}^{N} k_n.$$

Since θ is the mean of the Poisson random variable, the Maximum Likelihood estimator is, once again, what the Law of Large Numbers would, asymptotically, propose. A legitimate question which might now arise is whether the Law of Large Numbers is really all we need for estimating mean and variance. A simple reasoning shows that this is not the case. In fact, for the Poisson distribution, the Law of Large Numbers would suggest the variance approximation

$$\text{var}(K) \approx \frac{1}{N} \sum_{n=1}^{N} \left(k_n - \frac{1}{N}\sum_{j=1}^{N} k_j\right)^2,$$

which is different from the Maximum Likelihood estimate of the variance θ_{ML} derived above.

It is instructive to see what the Maximum Likelihood estimator would give if, instead of the Poisson distribution, we use its Gaussian approximation with mean and variance equal to θ, an approximation usually advocated for large values of θ. Writing

$$\prod_{n=1}^{N} \pi(k_n \mid \theta) \approx \left(\frac{1}{2\pi\theta}\right)^{N/2} \exp\left(-\frac{1}{2\theta} \sum_{n=1}^{N}(k_n - \theta)^2\right)$$

$$= \left(\frac{1}{2\pi}\right)^{N/2} \exp\left(-\frac{1}{2}\left[\frac{1}{\theta}\sum_{n=1}^{N}(k_n - \theta)^2 + N\log\theta\right]\right),$$

it follows that we can approximate θ_{ML} by minimizing

$$L(S \mid \theta) = \frac{1}{\theta} \sum_{n=1}^{N}(k_n - \theta)^2 + N\log\theta.$$

To solve this minimization problem we set

$$\frac{\partial}{\partial\theta} L(S \mid \theta) = -\frac{1}{\theta^2} \sum_{n=1}^{N}(k_n - \theta)^2 - \frac{2}{\theta} \sum_{n=1}^{N}(k_n - \theta) + \frac{N}{\theta} = 0,$$

or, equivalently,

$$-\sum_{n=1}^{N}(k_n - \theta)^2 - 2\sum_{n=1}^{N}\theta(k_n - \theta) + N\theta = N\theta^2 + N\theta - \sum_{n=1}^{N} k_n^2 = 0,$$

from which it follows that

$$\theta_{\mathrm{ML}} \approx \left(\frac{1}{4} + \frac{1}{N} \sum_{n=1}^{N} k_n^2 \right)^{1/2} - \frac{1}{2}. \tag{2.10}$$

Thus the approximate Maximum Likelihood estimate obtained from the Gaussian approximation differs from what the exact density gave.

The following example is a first step towards the interpretation of inverse problems in terms of statistical inference and highlights certain important characteristic features of inverse problems and Maximum Likelihood estimators.

EXAMPLE 2.5: Let X be a multivariate Gaussian random variable

$$X \sim \mathcal{N}(x_0, \Gamma), \tag{2.11}$$

where $x_0 \in \mathbb{R}^n$ is unknown and $\Gamma \in \mathbb{R}^{n \times n}$ is a known symmetric positive definite matrix. Furthermore, assume that x_0 depends on *hidden parameters* $z \in \mathbb{R}^k$ through a linear equation,

$$x_0 = Az, \quad A \in \mathbb{R}^{n \times k}, \quad z \in \mathbb{R}^k, \tag{2.12}$$

and that we want to estimate z from independent realizations of X. The model (2.12) can be interpreted as the observation model for z in the *ideal, noiseless case*. Since the world is less than ideal, in general we observe noisy copies of x_0. To rephrase the problem, we write an observation model

$$X = Az + E, \quad E \sim \mathcal{N}(0, \Gamma), \tag{2.13}$$

from which it follows that

$$\mathrm{E}\{X\} = Az + \mathrm{E}\{E\} = Az = x_0,$$

and

$$\mathrm{cov}(X) = \mathrm{E}\{(X - Az)(X - Az)^{\mathrm{T}}\} = \mathrm{E}\{EE^{\mathrm{T}}\} = \Gamma,$$

showing that the model (2.13) is in agreement with the assumption (2.11). The probability density of X, given z, is

$$\pi(x \mid z) = \frac{1}{(2\pi)^{n/2} \det(\Gamma)^{1/2}} \exp\left(-\frac{1}{2}(x - Az)^{\mathrm{T}} \Gamma^{-1}(x - Az) \right).$$

Given N independent observations of X,

$$\{x_1, \ldots, x_N\}, \quad x_j \in \mathbb{R}^n,$$

the likelihood function is

$$\prod_{j=1}^{N} \pi(x_j \mid z) \propto \exp\left(-\frac{1}{2}\sum_{j=1}^{N}(x_j - Az)^{\mathrm{T}}\Gamma^{-1}(x_j - Az)\right),$$

where we introduce the notation "$\propto$" to mean "equal up to a multiplicative constant of no interest". The likelihood function can be maximized by minimizing the negative of the log-likelihood function,

$$L(S \mid z) = \frac{1}{2}\sum_{j=1}^{N}(x_j - Az)^{\mathrm{T}}\Gamma^{-1}(x_j - Az)$$

$$= \frac{N}{2}z^{\mathrm{T}}\left[A^{\mathrm{T}}\Gamma^{-1}A\right]z - z^{\mathrm{T}}\left[A^{\mathrm{T}}\Gamma^{-1}\sum_{j=1}^{N}x_j\right] + \frac{1}{2}\sum_{j=1}^{N}x_j^{\mathrm{T}}\Gamma^{-1}x_j.$$

By computing the gradient of $L(S \mid z)$ with respect to z,

$$\nabla_z L(S \mid z) = N\left[A^{\mathrm{T}}\Gamma^{-1}A\right]z - A^{\mathrm{T}}\Gamma^{-1}\sum_{j=1}^{N}x_j, \tag{2.14}$$

and setting it equal to zero, we find that the Maximum Likelihood estimator z_{ML} is the solution of the linear system

$$\left[A^{\mathrm{T}}\Gamma^{-1}A\right]z = A^{\mathrm{T}}\Gamma^{-1}\bar{x},$$

where

$$\bar{x} = \frac{1}{N}\sum_{j=1}^{N}x_j,$$

provided that the solution exists. The existence of the solution of the linear system (2.14) depends on the properties of the matrix $A \in \mathbb{R}^{n \times k}$.
In the particular case of a *single* observation, $S = \{x\}$, the search for the Maximum Likelihood estimator reduces to the minimization of

$$L(x \mid z) = (x - Az)^{\mathrm{T}}\Gamma^{-1}(x - Az).$$

By using the eigenvalue decomposition of the covariance matrix,

$$\Gamma = UDU^{\mathrm{T}},$$

or, equivalently,

$$\Gamma^{-1} = W^{\mathrm{T}}W, \quad W = D^{-1/2}U^{\mathrm{T}},$$

we can express the likelihood in the form

$$L(x \mid z) = \|W(Az - x)\|^2.$$

Hence, the search for the Maximum Likelihood estimator is reduced to the solution of a *weighted least squares problem*[9] with weight matrix $W \in \mathbb{R}^{n \times n}$.

The last example shows that the maximum likelihood problem can be formulated naturally as a *basic problem of linear algebra*, namely as the search for the solution, in some sense, of a linear system of the type $Az = x$. This linear system does not necessarily have a solution, and if it does, the solution needs not be unique, or may be extremely sensitive to errors in the data. In the context of inverse problems, linear systems typically exhibit several of these problems, and the frequentist statistical framework, in general, provides no clue of how to overcome them. In this case, numerical analysis works for statistics, instead of statistics working for numerical analysis.

In the forthcoming chapters, we will investigate first what numerical linear algebra can do to overcome the challenges that arise. Later, we add the subjective aspect and see what statistical inference can indeed do for linear algebra.

Exercises

1. A classical example in dynamical systems and chaos theory is the iteration of the mapping

$$f : [0, 1] \to [0, 1], \quad x \mapsto 4x(1 - x).$$

Although fully deterministic, this function can be used to generate a sample that is chaotic; that is, which looks like a random sample, in the following way. Starting from a value $x_1 \in [0, 1]$, $x_1 \neq 0, 1/2, 1$, generate the sample $S = \{x_1, x_2, \ldots, x_N\}$ recursively by letting

$$x_{j+1} = f(x_j),$$

then calculate its mean and variance, and, using the Matlab function `hist`, investigate the behavior of the distribution as the sample size N increases.

2. Consider the Rayleigh distribution,

$$\pi(x) = \frac{x}{\sigma^2} \exp\left(-\frac{x^2}{2\sigma^2}\right), \quad x \geq 0,$$

discussed in Exercise 2 of Chapter 1. Given σ^2, generate a sample $S = \{x_1, x_2, \ldots, x_N\}$ from the Rayleigh distribution. If the sample had come

---

[9] One should probably use the term *generalized weighted least squares problem*, since usually, in the literature, the term *weighted least squares problem* refers to a problem where the matrix W is diagonal, i.e., each individual equation is weighted separately.

from a *log-normal distribution*, then the sample $S' = \{w_1, w_2, \ldots, w_N\}$, $w_j = \log x_j$, would be normally distributed. Write a Gaussian parametric distribution and estimate its mean and variance using the sample S. Investigate the validity of the log-normality assumption following the procedure of Example 2.1, that is, define the *credibility intervals* around the mean and count the number of sample points within the credibility intervals with different credibilities.

3. This exercise is meant to explore in more depth the meaning of the Maximum Likelihood estimator. Consider the Rayleigh distribution for which, unlike in the case of the normal distribution, the maximizer of $\pi(x)$ and the center of mass do not coincide. Calculate the maximizer and the center of mass, and check your result graphically.

Assuming that you have only *one* sample point, x_1, find the Maximum Likelihood estimator for σ^2. Does x_1 coincide with the maximizer or the mean?

3

The praise of ignorance: randomness as lack of information

"So you don't have a unique answer to your questions?"
"Adson, if I had, I would teach theology in Paris."
"Do they always have a right answer in Paris?"
"Never", said William, "but there they are quite confident of their errors."
(Umberto Eco: *The Name of the Rose*)

In the previous chapter the problem of statistical inference was considered from the frequentist's point of view: the data consist of a sample from a parametric probability density and the underlying parameters are *deterministic* quantities that we seek to estimate based on the data. In this section, we adopt the Bayesian point of view: *randomness simply means lack of information*. Therefore *any* quantity that is not known exactly is regarded as a random variable. The subjective part of this approach is clear: even if we believed that an underlying parameter corresponds to an existing physical quantity that could, in principle, be determined and therefore is conceptually a deterministic quantity, the lack of the *subject's* information about it justifies modeling it as a random variable. This is the general guiding principle that we will follow, applying it with various degrees of rigor[1].

When applying statistical techniques to inverse problems, the notion of *parameter* needs to be extended and elaborated. In classical statistics, parameters are often regarded as tools, like, for example, the mean or the variance, which identify a probability density. It is not uncommon that even in that context, parameters may have a physical[2] interpretation, yet they are treated as abstract parameters. In inverse problems, parameters are more often *by*

---

[1] After all, as tempting as it may sound, we don't want to end up teaching theology in Paris.

[2] The word "physical" in this context may be misleading in the sense that it makes us think of physics as a discipline. The use of the word here is more general, almost a synonym of "material" or "of observable nature", as opposed to something that is purely abstract.

definition physical quantities, but they *appear* as statistical model parameters defining probability densities. Disquisitions about such differences in interpretation may seem unimportant, but these very issues often complicate the dialogue between statisticians and "inversionists".

> EXAMPLE 3.1: Consider the general inverse problem of estimating a quantity $x \in \mathbb{R}^n$ that cannot be observed directly, but for which indirect observations of a related quantity $y \in \mathbb{R}^m$ are available. We may, for example, want to know the concentrations of certain chemical species (variable x) in a gas sample, but for some reason, we cannot measure them directly; instead, we observe spectral absorption lines of light that passes through the specimen (variable y). A mathematical model describing light absorption by a mixture of different chemical compounds ties these quantities together. The fact that the variables that we are interested in are concentration values already carries *a priori* the information that they cannot take on negative values. In addition, knowing where the sample is taken from, regardless of the subsequent measurement, we may have already a relatively good idea of what to expect to be found in the gas sample. In fact, the whole process of measuring may be performed to confirm a hypothesis about the concentrations.
>
> In order to set up the statistical framework we need to express the distribution of y in terms of the parameter x. This is done by constructing the *likelihood model*. The design of the prior model takes care of incorporating any available *prior* information.

As the preliminary example above suggests, the statistical model for inverse problems comprises two separate parts:

- The construction of the likelihood model;
- The construction of the prior model,

both of which make extensive use of *conditioning* and *marginalization*. When several random variables enter the construction of a model, by means of conditioning we can take into consideration one unknown at the time pretending that the others are given. This allows us to construct complicated models step by step. For example, if we want the joint density of x and y, we can write the density of y for fixed x and then deal with the density of x alone,

$$\pi(x, y) = \pi(y \mid x)\pi(x).$$

If, on the other hand, some of the variables appearing in the model are of no interest, we can eliminate them from the density by marginalizing them, that is by integrating them out. For example, if we have the joint density $\pi(x, y, v)$ but we are not interested in v, we can marginalize it as follows:

$$\pi(x, y) = \int \pi(x, y, v)dv.$$

The parameter v of no interest is often referred to as noise or as a nuisance parameter.

As statistics and probability are "common sense reduced to calculations", there is no universal prescription for the design of priors or likelihoods, although some recipes seems to be used more often than others. These will be discussed next.

3.1 Construction of Likelihood

In the previous section, the parametric likelihood density was viewed as a probability density from which, presumably, the observed data was generated. In the Bayesian context, the meaning of the likelihood is the same, with the only difference that parameters are seen as realizations of random variables. When discussing inverse problems, the unknown parameters always include the variables that we are primarily interested in. Hence, we can think of the likelihood as of the answers to the following question: *If we knew the unknown x and all other model parameters defining the data, how would the measurements be distributed?*

Since the construction of the likelihood starts from the assumption that, if x were known, the measurement y would be a random variable, it is important to understand the source of its randomness. Randomness being synonymous of lack of information, it suffices to analyze what makes the data deviate from the predictions of our observation model. The most common sources of deviations are

1. measurement noise in the data;
2. incompleteness of the observation model.

The probability density of the noise can, in turn, depend on unknown parameters, as will be demonstrated later in the examples. The second source of randomness is more complex, as it includes errors due to discretization, model reduction and more generally, all the shortcomings of a computational model, that is the discrepancy between the model and "reality" – in the heuristic sense of the word[3].

EXAMPLE 3.2: In inverse problems, it is very common to use additive models to account for measurement noise, as we did in the previous chapter. Assume that $x \in \mathbb{R}^n$ is the unknown of primary interest, that the observable quantity $y \in \mathbb{R}^m$ is ideally related to x through a functional dependence,

$$y = f(x), \quad f : \mathbb{R}^n \to \mathbb{R}^m, \tag{3.1}$$

---

[3] Writing a model for the discrepancy between the model and reality implicitly, and arrogantly, assumes that we know the reality and we are able to tell how the model fails to describe it. Therefore, the word "reality" is used in quotes, and should be understood as the "most comprehensive description available".

and that we are very certain of the validity of the model. The available measurement, however, is corrupted by noise, which we attribute to external sources or to instabilities in the measuring device, hence not dependent on x. Therefore we write the *additive noise model*,

$$Y = f(X) + E,$$

where $E : \Omega \to \mathbb{R}^m$ is the random variable modeling the noise. Observe that since X and Y are unknown, they are interpreted as random variables – hence upper case letters –, leading to a *stochastic extension* of the deterministic model (3.1).

Let us denote the distribution of the error by

$$E \sim \pi_{\text{noise}}(e).$$

Since we assume that the noise does not depend on X, fixing $X = x$ does not change the probability distribution of E. More precisely

$$\pi(e \mid x) = \pi(e) = \pi_{\text{noise}}(e).$$

If, on the other hand, X is fixed, the only randomness in Y is due to E. Therefore

$$\pi(y \mid x) = \pi_{\text{noise}}(y - f(x)), \tag{3.2}$$

that is, the randomness of the noise is translated by $f(x)$, as illustrated in Figure 3.1.

In this example we assume that the distribution of the noise is known. Although this is a common assumption, in practice the distribution of the noise is seldom known. More typically, the noise distribution itself depends on unknown parameters θ, thus

$$\pi_{\text{noise}}(e) = \pi_{\text{noise}}(e \mid \theta),$$

hence equation (3.2) becomes

$$\pi(y \mid x, \theta) = \pi_{\text{noise}}(y - f(x) \mid \theta).$$

To illustrate this, assume that the noise E is zero mean Gaussian with unknown variance σ^2,

$$E \sim \mathcal{N}(0, \sigma^2 I),$$

where $I \in \mathbb{R}^{m \times m}$ is the identity matrix. The corresponding likelihood model is then

$$\pi(y \mid x, \sigma^2) = \frac{1}{(2\pi)^{m/2} \sigma^m} \exp\left(-\frac{1}{2\sigma^2} \|y - f(x)\|^2\right),$$

with $\theta = \sigma^2$. If the noise variance is assumed known, we usually do not write the dependency explicitly, instead using the notation

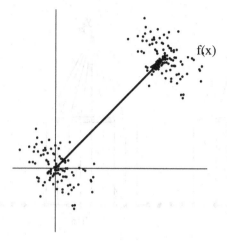

Fig. 3.1. Additive noise: the noise around the origin is shifted to a neighborhood of $f(x)$ without otherwise changing the distribution.

$$\pi(y \mid x) \propto \exp\left(-\frac{1}{2\sigma^2}\|y - f(x)\|^2\right),$$

hence ignoring the normalizing constant.

In the previous example we started from an ideal deterministic model and added independent noise. This is not necessarily always the case, as the following example shows: the forward model may be intrinsically probabilistic.

EXAMPLE 3.3: Assume that our measuring device consists of a collector lens and a counter of photons emitted from N sources, with average photon emission per observation time equal to x_j, $1 \le j \le N$, and that we want to estimate the total emission from each source over a fixed time interval. We take into account the geometry of the lens by assuming that the total photon count is the weighted sum of the individual contributions. When the device is above the jth source, it collects the photons from its immediate neighborhood. If the weights are denoted by a_k, with the index k measuring the offset to the left of the current position, the *expected* count is

$$\bar{y}_j = \mathrm{E}\{Y_j\} = \sum_{k=-L}^{L} a_k x_{j-k},$$

where the weights a_j are determined by the geometry of the lens and the index L is related to the width of the lens, as can be seen in Figure 3.2. Here it is understood that $x_j = 0$ if $j < 1$ or $j > N$.

Considering the ensemble of all source points at once, we can write

$$\bar{y} = \mathrm{E}\{Y\} = Ax,$$

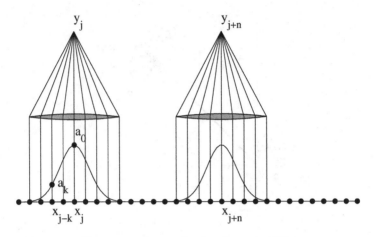

Fig. 3.2. The expected contribution from the different sources.

where $A \in \mathbb{R}^{N \times N}$ is the Toeplitz matrix

$$A = \begin{bmatrix} a_0 & a_{-1} & \cdots & a_{-L} & & & \\ a_1 & a_0 & & & \ddots & & \\ \vdots & & \ddots & & & a_{-L} & \\ a_L & & & \ddots & & & \vdots \\ & \ddots & & & a_0 & a_{-1} \\ & & a_L & \cdots & a_1 & a_0 \end{bmatrix}.$$

The parameter L defines the *bandwidth* of the matrix.

If the sources are weak, the observation model just described is a photon counting process. We may think that each Y_j is a Poisson process with mean $\overline{y}_j$,

$$Y_j \sim \text{Poisson}\big((Ax)_j\big),$$

that is,

$$\pi(y_j \mid x) = \frac{(Ax)_j^{y_j}}{y_j!} \exp\big(-(Ax)_j\big).$$

Observe that, in general, there is no guarantee that the expectations $(Ax)_j$ are integers. If we assume that consecutive measurements are independent, the random variable $Y \in \mathbb{R}^N$ has probability density

$$\pi(y \mid x) = \prod_{j=1}^{N} \pi(y_j \mid x) = \prod_{j=1}^{N} \frac{(Ax)_j^{y_j}}{y_j!} \exp\big(-(Ax)_j\big).$$

We express this relation simply by writing

$$Y \sim \text{Poisson}(Ax).$$

A model sometimes used in the literature assumes that the photon count is relatively large. In that case, using the Gaussian approximation of the Poisson density discussed in Chapter 1, the likelihood model becomes

$$\pi(y \mid x) \approx \prod_{j=1}^{N} \left(\frac{1}{2\pi(Ax)_j} \right)^{1/2} \exp\left(-\frac{1}{2(Ax)_j} \left(y_j - (Ax)_j \right)^2 \right) \tag{3.3}$$

$$= \left(\frac{1}{(2\pi)^L \det(\Gamma(x))} \right)^{1/2} \exp\left(-\frac{1}{2}(y - Ax)^{\mathrm{T}} \Gamma(x)^{-1}(y - Ax) \right),$$

where

$$\Gamma(x) = \text{diag}\big(Ax\big).$$

We could try to find an approximation of the Maximum Likelihood estimate of x by maximizing (3.3). However this maximization is nontrivial, since the matrix Γ appearing in the exponent and in the determinant is itself a function of x. In addition, a typical problem arising with convolution kernels is that the sensitivity of the problem to perturbations in the data is so high that *even if* we were able to compute the maximizer, it may be meaningless. Attempts to solve the problem in a stable way by means of numerical analysis have led to classical regularization techniques. The statistical approach advocated in this book makes an assessment of what we believe of a reasonable solution and incorporates this belief in the form of probability densities.

It is instructive to see what the Poisson noise looks like, and to see how Poisson noise differs from Gaussian noise with constant variance. In Figure 3.3, we have plotted a piecewise linear average signal $\bar{x}$ and calculated a realization of a Poisson process with $\bar{x}$ as its mean. It is evident that the higher the mean, the higher the variance, in agreement with the fact that the mean and the variance are equal. By visual inspection, Poisson noise could be confused with *multiplicative noise*, which will be discussed in the next example.

Before the example, let us point out a fact that is useful when composing probability densities. Assume that we have two random variables X and Y in $\mathbb{R}^n$ that are related via a formula

$$Y = f(X),$$

where f is a differentiable function, and that the probability distribution of Y is known. We write

$$\pi(y) = p(y),$$

to emphasize that the density of Y is given as a particular function p. It would be tempting to deduce that the density of X, denoted by $\pi(x)$, is obtained by substituting $y = f(x)$ in the function p. This, in general, is

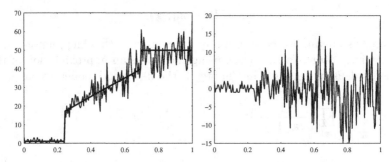

Fig. 3.3. Left panel: the average piecewise linear signal and a realization of a Poisson process, assuming that the values at each discretization point are mutually independent. Right panel: the difference between the noisy signal and the average.

not true, since probability densities represent *measures* rather than functions. The proper way to proceed is therefore to take the Jacobian of the coordinate transformation into consideration, by writing

$$\pi(y)dy = p(y)dy = p(f(x))|\det(Df(x))|dx,$$

where $Df \in \mathbb{R}^{n \times n}$ is the differential of f. Now we may identify the density of X as being

$$\pi(x) = p(f(x))|\det(Df(x))|. \tag{3.4}$$

In the following example, we make use of this formula.

EXAMPLE 3.4: Consider a noisy amplifier that takes in a signal $f(t)$ and sends it out amplified by a constant factor $\alpha > 1$. The ideal model for the output signal is

$$g(t) = \alpha f(t), \quad 0 \le t \le T.$$

In practice, however, it may happen that the amplification factor is not constant but fluctuates slightly around a mean value α_0. We write a discrete likelihood model for the output by first discretizing the signal. Let

$$x_j = f(t_j), \quad y_j = g(t_j), \quad 0 = t_1 < t_2 < \cdots < t_n = T.$$

Assuming that the amplification at $t = t_j$ is a_j, we have a discrete model

$$y_j = a_j x_j, \quad 1 \le j \le n,$$

and, replacing the unknown quantities by random variables, we obtain the stochastic extension

$$Y_j = A_j X_j, \quad 1 \le j \le n,$$

which we write in vector notation as

$$Y = A.X, \tag{3.5}$$

with the dot denoting componentwise multiplication of the vectors $A, X \in \mathbb{R}^n$. Assume that A as a random variable is independent of X, as is the case, for instance, if the random fluctuations in the amplification are due to thermal phenomena. If A has the probability density

$$A \sim \pi_{\text{noise}}(a),$$

to find the probability density of Y, conditioned on $X = x$, we fix X and write

$$A_j = \frac{Y_j}{x_j}, \quad 1 \leq j \leq n.$$

Applying formula (3.4), we obtain

$$\pi(y \mid x) = \frac{1}{x_1 x_2 \cdots x_n} \pi_{\text{noise}}\left(\frac{y.}{x}\right), \tag{3.6}$$

where the dot denotes that the division of the two vectors is componentwise.

Let us consider a special example where we assume that all the variables are positive, and A is *log-normally distributed*, i.e., the logarithm of A is normally distributed. For simplicity, we assume that the components of A are mutually independent, identically distributed, hence

$$W_i = \log A_i \sim \mathcal{N}(w_0, \sigma^2), \quad w_0 = \log \alpha_0.$$

To find an explicit formula for the density of A, we note that if $w = \log a$, where the logarithm is applied componentwise, we have

$$dw = \frac{1}{a_1 a_2 \cdots a_n} da,$$

thus the probability density of A is

$$\pi_{\text{noise}}(a) \propto \frac{1}{a_1 a_2 \cdots a_n} \exp\left(-\frac{1}{2\sigma^2} \|\log a - w_0\|^2\right)$$

$$= \frac{1}{a_1 a_2 \cdots a_n} \exp\left(-\frac{1}{2\sigma^2} \left\|\log\left(\frac{a.}{\alpha_0}\right)\right\|^2\right).$$

By substituting this formula in (3.6), we find that

$$\pi(y \mid x) \propto \frac{1}{y_1 y_2 \cdots y_n} \exp\left(-\frac{1}{2\sigma^2} \left\|\log\left(\frac{y.}{(\alpha_0 . x)}\right)\right\|^2\right).$$

Before moving onto the design of priors, let's look at an example with two different sources of noise.

EXAMPLE 3.5: In this example, we assume that the photon counting convolution device of Example 3.3 adds a noise component to the collected data. More precisely, we have an observation model of the form

$$Y = Z + E,$$

where $Z \sim \text{Poisson}(Ax)$ and $E \sim \mathcal{N}(0, \sigma^2 I)$.

To write the likelihood model, we begin with assuming that $X = x$ and $Z_j = z_j$ are known. Then

$$\pi(y_j \mid z_j, x) \propto \exp\left(-\frac{1}{2\sigma^2}(y_j - z_j)^2\right).$$

Observe that x does not appear explicitly here, but is, in fact, a hidden parameter that affects the distribution of Z_j. From the formula for the conditional probability density it follows that

$$\pi(y_j, z_j \mid x) = \pi(y_j \mid z_j, x)\pi(z_j \mid x), \quad \pi(z_j \mid x) = \frac{(Ax)_j^{z_j}}{z_j!}\exp\left(-(Ax)_j\right).$$

Since the value of z_j is not of interest here, we can marginalize it out. In view of the fact that z_j takes on only integer values greater than or equal to zero, we write:

$$\pi(y_j \mid x) = \sum_{z_j=0}^{\infty} \pi(y_j, z_j \mid x)$$

$$\propto \sum_{z_j=0}^{\infty} \pi(z_j \mid x)\exp\left(-\frac{1}{2\sigma^2}(y - z_j)^2\right),$$

which gives us the likelihood as a function of the variable of interest x. It is possible to replace the summation by using the Gaussian approximation for the Poisson distribution, but since the calculations are quite tedious and the final form is not so informative, we do not pursue that here.

3.2 Enter, Subject: Construction of Priors

The prior density expresses what we *know*[4], or more generally, *believe* about the unknown variable of interest prior to taking the measurements into account. The choice of a prior   accounts for the subjective portion of the

---

[4] The use of the word *knowing* could be replaced by *being certain of*, conforming with the notion of Wittgenstein on certainty: being certain of something does not imply that things are necessarily that way, which is the tragedy of science and the salvation of the art.

procedure. In fact, what we believe a priori about the parameters biases the search so as to favor solutions which adhere to our expectation. How significantly the prior is guiding our estimation of the unknowns depends in part on the information contents of the measurements: in the absence of good measured data, the prior provides a significant part of the missing information and therefore has a great influence on the estimate. If, on the other hand, the measured data is highly informative, it will play a big role in the estimation of the unknowns, leaving only a lesser role for the prior, unless we want to go deliberately against the observations. We actually use priors more extensively than we are generally ready to admit, and in variety of situations: often, when we claim not to know what to expect, our prior is so strong to rule out right away some of the possible outcomes as meaningless, and, in fact, the concept of meaningless only exists as a counterpart to meaningful, and its meaning is usually provided by a prior.

EXAMPLE 3.6: Here we give a trivial example of hidden priors: assume that you want somebody to pick "a number, any number". If that somebody is a normal[5] human being, the answer may be 3, 7, 13 – very popular choices in Judeo-Christian tradition – or 42 – a favorite number in the science fiction subculture[6]. You may think that you did not have any prior on what to expect, until you ask a space alien in disguise who, taking your question very seriously, starts to recite 40269167330954837..., continuing this litany of digits for over a week. Although eventually the stream of digits will end, as the number is finite after all, you start to wonder if a lifetime will be enough to finish a round of the game. Quite soon, probably much before the end of the week, you realize that this is not at all what you had in mind with "picking a number": the response is definitely against your prior. And in fact, you discover that not only your prior was there, but it was quite strong too: there are infinitely more numbers that would take more than a week – or a lifetime, for that matter – to say, than those you expected as an answer[7]!

Another, more pertinent example comes from the field of inverse problems.

---

[5] Preferably not from a math department, where people enjoy to come up with the most intricate answers for the simplest of questions, a tendency which becomes exacerbated when the answer involves numbers.

[6] In the subculture of mathematicians, even more exotic answers are likely, such as the Hardy–Ramanujan number, known also as the taxicab number, 1729. The point here is to notice that to obtain numbers no more complex than this, it takes some of the best number-minded brains. The human bias towards small numbers is enormous.

[7] Jose Luis Borges has an elegant short story, *The book of sand*, about a book with an endless number of pages. Opening the book and reading the page number would be tantamount to choosing "any" number. Borges, possibly aware of the problem, leaves the page numbering a mystery.

EXAMPLE 3.7: Your orthopedic surgeon has asked you to take along the MRI slides of your knee in the next visit, but, on your way out of the house, you accidentally took the envelope with the MRI slides of the brain. As soon as the slides are taken out of the envelope, the doctor knows that you have grabbed the wrong ones, in spite of having just told you not to have any idea what to expect to see in the slides.

The prior beliefs that we often have are *qualitative*, and it may be quite a challenge to translate them into *quantitative* terms. How do we express in a prior that we expect to see an MRI of the neck, or how do we describe quantitatively what radiologists mean by "habitus of a malignant tumor"? Clearly, we are now walking along the divide between art and science.

The obvious way to assign a prior is by stating that the unknown parameter of interest follows a certain distribution. Since the prior expresses our subjective belief about the unknown, our belief is a sufficient justification for its being. Note that the reluctance to choosing any specific prior, usually for fear of its subjective nature, *de facto* leads often to a claim that all values are equally likely, which, apart from its mathematical shortcomings, leaves it entirely up to the likelihood to determine the unknown. The results may be catastrophic.

The priors that we apply in day to day life are often inspired by our previous experience in similar situations. To see how we can proceed in the construction of a prior, we consider some examples.

EXAMPLE 3.7: Assume that we want to determine the level x of hemoglobin in blood by near-infrared (NIR) measurements at a patient's finger. If we have a collection of hemoglobin values measured directly from the patient's blood,

$$S = \{x_1, \ldots, x_N\},$$

we can think of them as *realizations* of a random variable X with an unknown distribution. As explained in Chapter 2, there are two possible ways for extracting information about the underlying distribution from S:

- The *non-parametric* approach looks at a histogram based on S and tries to infer what the underlying distribution is.
- The *parametric* approach proposes a parametric model, then computes the Maximum Likelihood estimate of the model parameters from the sample S.

Let us assume, for example, that a Gaussian model is proposed,

$$X \sim \mathcal{N}(x_0, \sigma^2).$$

We know from Chapter 2 that the Maximum Likelihood estimates for x_0 and σ^2 are,

$$x_{0,\text{ML}} = \frac{1}{N} \sum_{j=1}^{N} x_j,$$

and

$$\sigma_{\text{ML}}^2 = \frac{1}{N} \sum_{j=1}^{N} (x_j - x_{0,\text{ML}})^2,$$

respectively. We then assume that any future value x of the random variable is a realization from this Gaussian distribution. Thus, we postulate that:

- The unknown X is a random variable, whose probability distribution, called the *prior distribution*, is denoted by $\pi_{\text{prior}}(x)$,
- Guided by our prior experience, and assuming that a Gaussian prior is justifiable, we use the parametric model

$$\pi_{\text{prior}}(x) = \frac{1}{\sqrt{2\pi\sigma^2}} \exp\left(-\frac{1}{2\sigma^2}(x - x_0)^2\right),$$

with x_0 and σ^2 determined experimentally from S by the formulas above.

Methods in which prior parameters are estimated empirically, either from previous observations or simultaneously with the unknown from the current data, are called *empirical Bayes methods*.

The prior may be partly based on a physical, chemical or biological model, as in the following example.

EXAMPLE 3.8: Consider a petri dish with a culture of bacteria whose density we want to estimate. For simplicity, let's assume that we have a rectangular array of squares, where each square contains a certain number of bacteria, as illustrated in Figure 3.4, and that we are interested in estimating the density of the bacteria from some indirect measurements[8]. We begin with setting up a model based on our belief about bacterial growth. For example, we may assume that the number of bacteria in a square is approximately the average of bacteria in the neighboring squares,

$$x_j \approx \frac{1}{4}(x_{\text{left},j} + x_{\text{right},j} + x_{\text{up},j} + x_{\text{down},j}), \qquad (3.7)$$

see Figure 3.4 for an explanation. Since the squares at the boundary of the petri dish have no neighbors in some directions, we need to modify (3.7) to account for their special status. The way in which we will handle this turns out to be related to the choice of boundary conditions for partial differential equations. For example, we can assume that $x_j = 0$ in pixels outside the square.

Let N be the number of pixels and $A \in \mathbb{R}^{N \times N}$ be the matrix with jth row

---

[8] Aside from the fact that direct count of bacteria would be impossible, it would amount to moments of unmatched enjoyment.

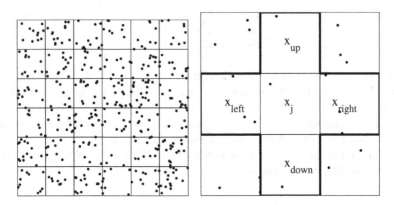

Fig. 3.4. Square array of bacteria. On the right we zoom in on a neighborhood system.

$$\begin{array}{cccc} \text{(up)} & \text{(down)} & \text{(left)} & \text{(right)} \end{array}$$
$$A(j,:) = \begin{bmatrix} 0 \cdots 1/4 \cdots & 1/4 \cdots & 1/4 \cdots & 1/4 \cdots 0 \end{bmatrix},$$

with the understanding that for boundary and corner pixels, some of the columns may be missing, corresponding to the assumption that no contribution from the outside of the array is coming.

If we were absolutely certain of the validity of the model (3.7), we would replace the approximate sign with strict equality. In other words we would assume that

$$x = Ax. \tag{3.8}$$

which clearly does not work! In fact, after rewriting (3.8) as

$$(I - A)x = 0,$$

it immediately follows that $x = 0$, since $I - A$ is an invertible matrix. Therefore, to relax the model (3.8) by admitting some uncertainty, we define X to be a random variable and write a stochastic model

$$X = AX + R, \tag{3.9}$$

where R expresses the uncertainty of the averaging model. The lack of information about the values of R is encoded in its probability density,

$$R \sim \pi_{\text{mod.error}}(r),$$

and, by writing $R = X - AX$, we define that the prior density for X is

$$\pi_{\text{prior}}(x) \propto \pi_{\text{mod.error}}(x - Ax).$$

Equation (3.9) defines an *autoregressive Markov model*, and in this context R is referred to as an *innovation process*. In particular, if R is Gaussian with mutually independent, identically distributed components,

$$R \sim \mathcal{N}(0, \sigma^2 I),$$

the prior for X is of the form

$$\pi_{\text{prior}}(x \mid \sigma^2) = \left(\frac{1}{2\pi\sigma^2}\right)^{N/2} \exp\left(-\frac{1}{2\sigma^2}\|x - Ax\|^2\right)$$

$$= \left(\frac{1}{2\pi\sigma^2}\right)^{N/2} \exp\left(-\frac{1}{2\sigma^2}\|Lx\|^2\right),$$

where

$$L = I - A.$$

We remark that if we have no natural way of fixing the value of σ^2, estimating it on the basis of observations is part of the larger estimation problem. One possible line of thought for choosing the variance σ^2 could be as follows: suppose that you do expect to find about M bacteria in the whole dish. That means that, in the average, one square contains about $m = M/N$ bacteria. The fluctuation around the postulated average value in a single square is probably not much more than m, so a reasonable choice could be $\sigma \approx m$.

We conclude this example by showing a Matlab code to construct the matrix A. Since this matrix is sparse, it is advisable to construct it as a sparse matrix. Sparse matrices are defined in Matlab by three vectors of equal length. The first one contains the indices of the rows with non-vanishing entries, the second one the column indices and the third one the actual non-vanishing values. These vectors are called rows, cols and vals in the code below.

```
n = 50;  % Number of pixels per directions

% Creating an index matrix to enumerate the pixels

I = reshape([1:n^2],n,n);

% Right neighbors

Icurr = I(:,1:n-1);
Ineigh = I(:,2:n);
rows = Icurr(:);
cols = Ineigh(:);
vals = ones(n*(n-1),1);

% Left neighbors

Icurr = I(:,2:n);
Ineigh = I(:,1:n-1);
```

```
rows = [rows;Icurr(:)];
cols = [cols;Ineigh(:)];
vals = [vals;ones(n*(n-1),1)];

% Upper neighbors

Icurr = I(2:n-1,:);
Ineigh = I(1:n-1,:);
rows = [rows;Icurr(:)];
cols = [cols;Ineigh(:)];
vals = [vals;ones(n*(n-1),1)];

% Lower neighbors

Icurr = I(1:n-1,:);
Ineigh = I(2:n,:);
rows = [rows;Icurr(:)];
cols = [cols;Ineigh(:)];
vals = [vals;ones(n*(n-1),1)];

A = 1/4*sparse(rows,cols,vals);
L = speye(n^2) - A;
```

Observe that L is, in fact, a second order finite difference matrix built from the mask

$$\begin{bmatrix} & -1/4 & \\ -1/4 & 1 & -1/4 \\ & -1/4 & \end{bmatrix},$$

hence L is, up to a scaling constant, a discrete approximation of the Laplacian

$$-\Delta = -\frac{\partial^2}{\partial x_1^2} - \frac{\partial^2}{\partial x_2^2}$$

in the unit square. Indeed, if

$$x_j = f(p_j),$$

where p_j is a point in the jth pixel, the finite difference approximation of the Laplacian of f can be written in the form

$$-\Delta f(p_j) \approx \frac{4}{h^2}(Lx)_j,$$

where h is the length of the discretization step. At the boundaries, we assume that f extends smoothly by zero outside the square.

The prior model derived in the last example corresponds to what is often referred to as *second order smoothness prior*. The reason for the name is that this prior assigns a higher probability to vectors $x \in \mathbb{R}^N$ corresponding to discrete approximations of functions with a small second derivative, since the exponential is larger when the negative exponent is smaller, that is, when Lx has a small norm. This is the case, in general, for vectors x which are discretizations of smooth functions. We remark that a Gaussian smoothness prior does not exclude the occurrence of large jumps between adjacent pixels, but it gives them an extremely low probability.

Before leaving, temporarily, the theme of prior distributions, we want to report on a typical discussion which could occur after a presentation of Bayesian solutions of inverse problems on how to estimate a gray scale image from a blurred and noisy copy. In the image processing literature, this is referred to as a denoising and deblurring problem. The image can be represented as a pixel matrix, each entry assigning a gray scale value to the corresponding pixel. Assume that, after stacking the pixel values in one long vector, a Gaussian prior has been constructed for the image vectors. A typical question is: "Can you really assume that the prior density is Gaussian[9]?" While at first sight such a question may seem reasonable, once properly analyzed, it becomes rather obscure. In fact, the question is based on the Platonic view that *there is a true prior*. But, true prior of what, we may ask. All possible images, maybe? Most certainly not, and in fact *all possible images* is a useless category, as is the category of all possible realizations of all random process, or all possible worlds. Maybe the set should be restricted to all possible images that can be realized in the particular application that we have in mind? But such category is also useless, unless we specify its genesis by describing, for instance, the probability distribution of the images. And at this point we face the classic problem of which came first, the hen or the egg. A more reasonable question would therefore be, what the given prior implies. As we shall see, a possible way of exploring the implications of a prior distribution is to use sampling techniques.

3.3 Posterior Densities as Solutions of Statistical Inverse Problems

Within the previous sections of this chapter we introduced the main actors in our Bayesian play, the prior and the likelihood. From now on, the marginal density of the unknown of primary interest will be identified as prior density, and denoted by π_{prior}. Notice that both the likelihood and the prior may contain parameters whose values we are not very confident about; the natural Bayesian solution is to regard them as random variables, too.

If we let X denote the random variable to be estimated and Y the random variable that we observe, from the *Bayes formula* we have that

---

[9] Note that when saying that the prior is Gaussian, we do not intend a prior with independent equally distributed components, so the obvious arguments referring to non-negativity of the image do not necessarily apply here.

$$\pi(x \mid y) = \frac{\pi_{\text{prior}}(x)\pi(y \mid x)}{\pi(y)}, \quad y = y_{\text{observed}}. \tag{3.10}$$

The conditional density $\pi(x \mid y)$, called the *posterior density*, expresses the probability of the unknown given that the observed parameters take on the values given as the data of the problem and our prior belief. In a Bayesian statistical framework, *the posterior density is the solution of the inverse problem*.

Two questions are now pertinent. The first one is how to find the posterior density and the second how to extract information from it in a form suitable for our application. The answer to the first question has already been given: from the prior and the likelihood, the posterior can be assembled via the Bayes formula. The answer to the second question will be given in the remaining chapters. As a prelude to the ensuing discussion, let us consider the following example.

EXAMPLE 3.9: Consider a linear system of equations with noisy data,

$$y = Ax + e, \quad x \in \mathbb{R}^n, \ y, e \in \mathbb{R}^m, \ A \in \mathbb{R}^{m \times n},$$

and let

$$Y = AX + E$$

be its stochastic extension, where X, Y and E are random variables. A very common assumption is that X and E are independent and Gaussian,

$$X \sim \mathcal{N}(0, \gamma^2 \Gamma), \quad E \sim \mathcal{N}(0, \sigma^2 I),$$

where we have assumed that both random variables X and E have zero mean. If this is not the case, the means can be subtracted from the random variable, and we arrive at the current case. The covariance of the noise indicates that each component of Y is contaminated by independent identically distributed random noise. We assume that the matrix Γ in the prior is known. The role of the scaling factor γ will be discussed later. The prior density is therefore of the form

$$\pi_{\text{prior}}(x) \propto \exp\left(-\frac{1}{2\gamma^2}x^{\mathrm{T}}\Gamma^{-1}x\right),$$

The likelihood density, assuming that the noise level σ^2 is known, is

$$\pi(y \mid x) \propto \exp\left(-\frac{1}{2\sigma^2}\|y - Ax\|^2\right).$$

It follows from the Bayes formula that the posterior density is

$$\pi(x \mid y) \propto \pi_{\text{prior}}(x)\pi(y \mid x)$$

$$\propto \exp\left(-\frac{1}{2\gamma^2}x^{\mathrm{T}}\Gamma^{-1}x - \frac{1}{2\sigma^2}\|y - Ax\|^2\right)$$

$$= \exp\left(-V(x \mid y)\right),$$

where

$$V(x \mid y) = \frac{1}{\gamma^2} x^{\mathrm{T}} \Gamma^{-1} x + \frac{1}{2\sigma^2} \|y - Ax\|^2.$$

Since the matrix Γ is symmetric positive definite, so is its inverse, thus admitting a symmetric, e.g., Cholesky, factorization:

$$\Gamma^{-1} = R^{\mathrm{T}} R.$$

With this notation,

$$x^{\mathrm{T}} \Gamma^{-1} x = x^{\mathrm{T}} R^{\mathrm{T}} R x = \|Rx\|^2,$$

and we define

$$T(x) = 2\sigma^2 V(x \mid y) = \|y - Ax\|^2 + \delta^2 \|Rx\|^2, \quad \delta = \frac{\sigma}{\gamma}. \tag{3.11}$$

This functional T, sometimes referred to as the *Tikhonov functional*, plays a central role in the classical regularization theory.

The analogue of the Maximum Likelihood estimator in the Bayesian setting maximizes the posterior probability of the unknowns and is referred to as the *Maximum A Posteriori (MAP) estimator*:

$$x_{\mathrm{MAP}} = \arg\max \pi(x \mid y).$$

Note that

$$x_{\mathrm{MAP}} = \arg\min V(x \mid y), \quad V(x \mid y) = -\log \pi(x \mid y).$$

In this particular case we have that

$$x_{\mathrm{MAP}} = \arg\min \left(\|y - Ax\|^2 + \delta^2 \|Rx\|^2 \right). \tag{3.12}$$

When the posterior density is Gaussian, the Maximum A Posteriori estimate coincides with the *Conditional Mean* (CM), or *Posterior Mean* estimate,

$$x_{\mathrm{CM}} = \int x \pi(x \mid y) dy.$$

It is immediate to check that if we have $R = I$ and let γ increase without bounds, the parameter δ goes to zero and the Maximum A Posteriori estimator reduces formally to the Maximum Likelihood estimator, namely the estimation of x is entirely based on the likelihood density.

In our discussion of the Maximum Likelihood estimation in Chapter 2, Example 2.4, we saw that its calculation can be reduced to solving a system of linear equations

$$Ax = y, \tag{3.13}$$

possibly in the least squares sense. It was also pointed out that even when the matrix A is square, if it is numerically singular the calculation of the Maximum Likelihood estimate becomes problematic. Classical regularization techniques are based on the idea that an ill-posed problem is replaced by a nearby problem that is well-posed. *Tikhonov regularization*, for example, replaces the linear system (4.1) with the minimization problem

$$\min \left\{ \|y - Ax\|^2 + \delta^2 \|Rx\|^2 \right\}, \tag{3.14}$$

where the first term controls the fidelity of the solution to the data and the second one acts as a *penalty*. The matrix R is typically selected so that large $\|Rx\|$ corresponds to an undesirable feature of the solution. The role of the penalty is to keep this feature from growing unnecessarily. A central problem in Tikhonov regularization is how to choose the value of the parameter δ judiciously. A Tikhonov regularized solution is the result of a compromise between fitting the data and eliminating unwanted features. As we have seen, this fits well into the Bayesian framework, since the penalty can be given a statistical interpretation via the prior density.

A question which comes up naturally at this point is if and how Tikhonov regularization, or MAP estimation, can avoid the numerical problems that make Maximum Likelihood estimation generally infeasible for noisy data. The answer can be found by writing the functional to be minimized in (3.14) in the form

$$\|y - Ax\|^2 + \delta^2 \|Rx\|^2 = \left\| \begin{bmatrix} y \\ 0 \end{bmatrix} - \begin{bmatrix} A \\ \delta R \end{bmatrix} x \right\|^2,$$

which reveals that the Maximum A Posteriori estimate is the least squares solution of the linear system

$$\begin{bmatrix} A \\ \delta R \end{bmatrix} x = \begin{bmatrix} y \\ 0 \end{bmatrix}.$$

While the original problem (3.13) may be ill-posed, the augmentation of the matrix A by δR considerably reduces the ill-posedness of the problem. The quality of the solution depends on the properties of the matrix R.

Exercises

1. A patient with a lump in the breast undergoes a mammography. The radiologist who examines her mammogram believes that the lump is malignant, that is, the result of the mammogram is positive for malignancy. The radiologist's record of true and false positives is shown in the table below:

	Malignant	Benign
Positive mammogram	0.8	0.1
Negative mammogram	0.2	0.9

The patient, with the help of the internet, finds out that in the whole population of females, the probability of having a malignant breast tumor at her age is 0.5 percent. Without thinking much further, she takes this number as her prior belief of having cancer.

(a) What is the probability of the mammogram result being positive for malignancy?

(b) What is the conditional probability of her having a malignant tumor, considering the fact that the mammogram's result was positive?

(c) What problems are there with her selection of the prior, others than the possible unreliability of internet?

2. A person claims to be able to guess a number from 1 to 10 which you are asked to think of. He has a record of having succeeded 8 times out of 10. You question the claim and in your mind assign a certain probability x of him having indeed such gift, but decide to give the poor devil a chance and let him try to guess the number between 1 and 10 that you are thinking. He guesses correctly, but you are still not convinced. In other words, even after his successful demonstration, you still think that he is a swindler and that the probability of such an extraordinary gift is less than 0.5. How low must your prior belief x have been for this to happen?

3. Rederive the result of Example 3.4 in a slightly different way: By taking the logarithm of both sides of equation (3.5), we obtain

$$\log Y = \log X + \log A = \log X + W,$$

where W is normally distributed. Write the likelihood density for $\log Y$ conditioned on $X = x$ and derive the likelihood for Y.

4

Basic problem in numerical linear algebra

We have seen in the previous chapter that the computation of both the Maximum Likelihood and the Maximum A Posteriori estimators often requires solving a linear system of equations. In this chapter we review methods for the efficient solution of linear systems of equations in the classical or in the least squares sense, and we give some insight into the complications that might arise. In this chapter we review methods of traditional linear algebra for solving ill-conditioned problems, the purpose being to prepare the way for statistical methods to solve problems in numerical linear algebra. The Trojan horse that makes this possible is the Bayesian framework, prior and likelihood as its head and tail, bringing into the numerical algorithms information that would not be otherwise usable.

4.1 What is a solution?

The main problem of numerical linear algebra that is of concern to us is how to compute the solution of a linear system

$$Ax = b \qquad A \in \mathbb{R}^{m \times n}, \quad x \in \mathbb{R}^n, \quad b \in \mathbb{R}^m. \qquad (4.1)$$

Since the linear system (4.1) is not necessarily square, and, even when square, not necessarily solvable, the word "solution" may have different meanings depending on the values of m, n and on the properties of A and b. In particular,

1. If $n > m$, the problem is *underdetermined*, i.e., there are more unknowns than linear equations to determine them;
2. If $n = m$, the problem is *formally determined*, i.e., the number of unknowns is equal to the number of equations;
3. If $n < m$, the problem is *overdetermined*, i.e., the number of equations to satisfy exceeds that of the unknowns.

We now discuss the usual meaning of the word "solution" in each of these three cases, a graphical illustration of which can be seen in Figure 4.1.

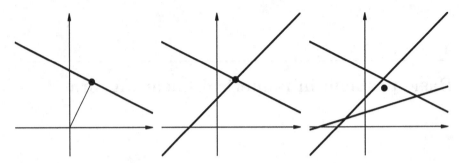

Fig. 4.1. Meaning of a solution. Left: minimum norm solution of an underdetermined system ($n = 2$, $m = 1$). Middle: exact solution of a non-singular system ($n = m = 2$). Right: least squares solution of an overdetermined system ($n = 2$, $m = 3$).

If $m < n$, the number of unknowns exceeds the number of equations. Since, in general, we can expect that at most as many unknowns as linear equations can be determined, at least $m - n$ unknowns cannot be determined without additional information about the solution. One way to solve this underdetermined linear system is to first fix the values of $m - n$ of the unknowns, then to determine the remaining ones as functions of the assigned values. Another common procedure for the solution of underdetermined linear systems is to look for a *minimum norm solution*, intended as a solution vector of minimal Euclidean norm,

$$x_{\mathrm{MN}} = \arg\min\left\{\|x\| \mid Ax = b\right\}.$$

If $m = n$, we have as many linear equations as unknowns, but this only guarantees that a solution exists and is unique if the matrix A is nonsingular. If A is not invertible and the system is consistent, some of the equations are redundant, in the sense that they are linear combinations of the remaining ones, or the system may be inconsistent. In the first case the system is effectively underdetermined.

If $m > n$, the number of linear equations exceeds the number of unknowns and, in general, we cannot expect to find a vector x such that $Ax = b$. In general, we seek a vector x such that

$$x_{\mathrm{LS}} = \arg\min\|b - Ax\|^2, \tag{4.2}$$

where $\|\cdot\|$ is the Euclidean vector norm. The minimizer, if it exists, is called the *least squares solution* of (4.1) because it minimizes the sum of the squares of the components of the residual error

$$r = b - Ax.$$

Algorithms for solving overdetermined systems are discussed below.

The choice of the best algorithm for solving (4.1) typically depends on several factors, with the size of the problem playing an important role. In general, we distinguish algorithms for solving linear systems of equations into two broader classes: *direct methods* and *iterative methods*.

4.2 Direct linear system solvers

Direct methods solve (4.1) by first factorizing the matrix A and then solving simpler linear systems. Since typically the factorization of the matrix A requires

- The matrix A to be explicitly available;
- Sufficient memory allocation for the matrix A and its factors;
- A number of floating point operations proportional to the cube of the dimension of the problem;

they are most suitable for linear systems of small or medium dimensions. Furthermore, direct methods are of the "all or nothing" kind, in the sense that if the process is not completed, no partial information about the solution is available. Although we shall not present here a systematic treatment of direct methods, which can be found in any linear algebra textbook and is not a central topic for these notes, the main algorithmic points are summarized below in the form of examples.

EXAMPLE 4.1: If (4.1) is a square system, i.e., $m = n$, a standard procedure is to perform a decomposition of the matrix A into a product of triangular matrices and to solve the resulting subproblems by forward and back substitution. If A is nonsingular, the algorithm referred to as Gaussian elimination with partial pivoting produces a factorization of the form

$$PA = LU$$

where L and U are the lower triangular and the upper triangular matrices,

$$L = \begin{bmatrix} 1 & & & \\ * & 1 & & \\ \vdots & & \ddots & \\ * & \cdots & * & 1 \end{bmatrix}, \quad U = \begin{bmatrix} * & * & \cdots & * \\ & * & & * \\ & & \ddots & \vdots \\ & & & * \end{bmatrix},$$

where the asterisks are placeholders for possibly non-zero elements and P is a permutation matrix. Having computed the LU-decomposition of A, we then solve the two triangular systems

$$Ly = Pb, \qquad Ux = y.$$

If the matrix A is symmetric positive definite, the linear system (4.1) can be solved by first computing the Cholesky factorization of A:

$$A = R^{\mathrm{T}}R$$

where R is upper triangular, then solving

$$R^T y = b, \quad Rx = y.$$

Since the Cholesky factorization is only defined for symmetric positive definite matrices, this solution method, which requires only half of the storage and of the computations for the matrix factorization, can only be applied to a restricted class of linear systems. When attempting to compute the Cholesky factorization of a symmetric matrix that is not positive definite, the algorithm breaks down while trying to compute the square root of a negative number. It is important to remember that matrices which are mathematically symmetric positive definite may fail to be numerically so. This is true in particular when the matrix is of large dimensions and it arises from the discretization of an ill-posed problem. The errors accumulated during the process of computing the Cholesky factorization may make the matrix numerically singular or indefinite, thus causing the algorithm to break down prior to completion.

Let's now discuss how to solve overdetermined linear systems of equations.

EXAMPLE 4.2: Consider an overdetermined linear system

$$Ax = b, \quad A = \begin{bmatrix} * & \cdots & * \\ * & \cdots & * \\ \vdots & & \vdots \\ * & \cdots & * \\ * & \cdots & * \end{bmatrix} \in \mathbb{R}^{m \times n}, \quad m > n.$$

If the columns of A are linearly independent, the minimizer of (4.2) exists and is unique. Multiplying both sides of the above equation by A^T from the left, we obtain the *normal equations*,

$$A^T A x = A^T b. \tag{4.3}$$

This is a square nonsingular linear system, whose coefficient matrix $A^T A \in \mathbb{R}^{n \times n}$ is symmetric and positive definite. In fact,

$$x^T A^T A x = (Ax)^T Ax = \|Ax\|^2 \geq 0,$$

and this quadratic form is zero only if $Ax = 0$. But since Ax is a linear combination of the columns of A, which by assumption are linearly independent, $Ax = 0$ only if $x = 0$. The normal equation can be solved by using the Cholesky factorization.

If the columns of A are linearly independent, the solution of the normal equation (4.3) is the least squares solution,

$$x_{LS} = (A^T A)^{-1} A^T b = A^\dagger b.$$

The matrix $A^\dagger \in \mathbb{R}^{n \times m}$ is called the *pseudoinverse* of A. It is easy to check that if A is square and invertible, the pseudoinverse coincides with the usual inverse.

Forming and solving the normal equations is not necessarily the best way to compute the solution of a least squares problem, in particular when the coefficient matrix is ill-conditioned and the right-hand side is contaminated by errors. These issues will be discussed later in this chapter. For now, we will just remark that the compression of the information contained in the matrix A occurring when forming $A^T A$ is likely to introduce into the system additional errors which can compromise the accuracy of the solution.

In the case of ill-conditioned least squares problems it might be preferable to use the QR factorization of the matrix A,

$$A = QR,$$

where $Q \in \mathbb{R}^{m \times m}$ is an orthogonal matrix, that is, $Q^{-1} = Q^T$, and R is upper triangular,

$$R = \begin{bmatrix} * & \cdots & * \\ & \ddots & \vdots \\ & & * \\ & & \\ & & \end{bmatrix} = \begin{bmatrix} R_1 \\ 0 \end{bmatrix} \in \mathbb{R}^{m \times n},$$

with $R_1 \in \mathbb{R}^{n \times n}$. Multiplying both sides of the linear system from the left by the transpose of Q and replacing A by its QR factorization, we arrive at the linear triangular system

$$Rx = Q^T b,$$

which can be partitioned as

$$\begin{bmatrix} R_1 x \\ 0 \end{bmatrix} = \begin{bmatrix} (Q^T b)_1 \\ (Q^T b)_2 \end{bmatrix}.$$

Clearly, the last $m - n$ equations are either satisfied or not, depending on the right hand side b; the variable $x \in \mathbb{R}^n$ plays no role in that. The first n equations can be solved by backsubstitution, if the diagonal elements of R_1 are non-zero.

The pseudoinverse is useful also when analyzing underdetermined linear systems and, as a special case, formally determined linear systems with non-invertible square matrix, as we will see in the following example.

EXAMPLE 4.3: A useful tool for analyzing linear systems is the *singular value decomposition* (SVD) of the matrix A,

$$A = UDV^{\mathrm{T}}, \quad A \in \mathbb{R}^{m \times n}.$$

Here $U \in \mathbb{R}^{m \times m}$ and $V \in \mathbb{R}^{n \times n}$ are orthogonal matrices, that is, $U^{-1} = U^{\mathrm{T}}$, $V^{-1} = V^{\mathrm{T}}$ and $D \in \mathbb{R}^{m \times n}$ is a diagonal matrix of the form

$$D = \begin{bmatrix} d_1 & & & \\ & d_2 & & \\ & & \ddots & \\ & & & d_n \\ & & & \\ & & & \end{bmatrix}, \quad \text{if } m > n,$$

or

$$D = \begin{bmatrix} d_1 & & & & \\ & d_2 & & & \\ & & \ddots & & \\ & & & d_m & \end{bmatrix}, \quad \text{if } m < n,$$

or, briefly,

$$D = \mathrm{diag}\big(d_1, d_2, \ldots, d_{\min(m,n)}\big), \quad d_j \geq 0.$$

We may assume that the singular values $d_1, d_2, \ldots$ are in decreasing order. Using the singular value decomposition it is possible to transform any linear system into a diagonal one by two changes of variables,

$$Dx' = b', \quad x' = V^{\mathrm{T}}x, \quad b' = U^{\mathrm{T}}b. \tag{4.4}$$

In the overdetermined case, the diagonal linear system is of the form

$$b'_j = \begin{cases} d_j x'_j, & 1 \leq j \leq n, \\ 0, & n < j \leq m. \end{cases}$$

Whether the $m - n$ equations are satisfied or not depends entirely of the right hand side b. Whether we can satisfy the first n equations, on the other hand, depends on the singular values d_j. Assume that the first p singular values are positive, and the rest of them vanish, i.e.,

$$d_1 \geq d_2 \geq \cdots \geq d_p > d_{p+1} = d_{p+2} = \cdots d_n = 0.$$

Clearly the only possible choice for the first p components of the solution vector x is

$$x'_j = \frac{b'_j}{d_j}, \quad 1 \leq j \leq p,$$

while the remaining components of x' can be chosen arbitrarily. In the absence of additional information about the desired solution, it is customary, adhering to the minimum norm solution paradigm, to set them to zero. This is mathematically equivalent to defining the solution of the linear system to be

$$x = A^\dagger b,$$

where

$$A^\dagger = V D^\dagger U^\mathrm{T}, \quad D^\dagger = \mathrm{diag}\left(\frac{1}{d_1}, \dots, \frac{1}{d_p}, 0, \dots, 0\right) \in \mathbb{R}^{m \times n}. \qquad (4.5)$$

Here we use the pseudoinverse notation because when the columns of A are linearly independent, (4.5) is equivalent to the definition of pseudoinverse. The definition (4.5), however, is more general: in the case where $m < n$, it gives the minimum norm solution.

Although from the theoretical point of view the singular value decomposition is a very versatile tool to analyze linear systems, it can only be used for problems of small dimensions, because of its elevated computational cost. The singular value decomposition will be revisited when the ill–conditioning of linear systems is discussed.

4.3 Iterative linear system solvers

Many of the applications that we have in mind give rise to linear systems of large dimensions, whose solution requires the use of iterative methods. Iterative linear system solvers become of interest when

- the dimensions of the linear system are very large;
- the matrix A is not explicitly given;
- we only want an approximate solution of the linear system.

The second case is typical in applications where the *action* of a linear mapping is easy to calculate, i.e, for any given vector x, the output of $x \mapsto Ax$ is easily available. Examples of such matrix-free applications can be found in signal and image processing. The following example illustrates the use of Fast Fourier Transform (FFT) for computing convolution integrals without forming convolution matrices.

EXAMPLE 4.4: Consider a signal, $f(t)$, $0 \le t \le T$, that has been sampled at discrete time instances,

$$x_j = f(t_j), \quad t_j = j\frac{T}{n}, \quad 0 \le j \le n - 1.$$

Its convolution with a kernel h is defined as an integral

$$g(t) = \int_0^T h(t - s)f(s)ds, \quad 0 \le t < T,$$

which can be discretized and evaluated as a matrix product similarly as in Example 3.3. An alternative way is to use Fourier transformation. Working with Fourier series of periodic functions, let us extend the input signal to a T-periodic function, f_T, and modify the convolution operator accordingly,

$$g(t) = \int_{-\infty}^{\infty} h(t - s)f_T(s)ds$$

$$= \int_{-\infty}^{\infty} h(s)f_T(t - s)ds, \quad 0 \le t < T,$$

so that the output g is also T-periodic. The kth Fourier coefficient of the output signal,

$$\widehat{g}(k) = \frac{1}{T}\int_0^T \exp\left(-i\frac{2\pi}{T}tk\right)g(t)dt$$

$$= \int_{-\infty}^{\infty} h(s)\left[\frac{1}{T}\int_0^T \exp\left(-i\frac{2\pi}{T}(t-s)k\right)f_T(t-s)dt\right]\exp\left(-i\frac{2\pi}{T}sk\right)ds$$

$$= \widehat{f}(k)\int_{-\infty}^{\infty} h(s)\exp\left(-i\frac{2\pi}{T}sk\right)ds$$

$$= \widehat{h}(k)\widehat{f}(k),$$

is then the product of the kth Fourier coefficient of f and the Fourier transform of h at k,

$$\widehat{h}(k) = \int_{-\infty}^{\infty} h(s)\exp\left(-i\frac{2\pi}{T}sk\right)ds.$$

The convolved output signal is then expressed as a Fourier series,

$$g(t) = \sum_{k=-\infty}^{\infty} \widehat{g}(k)\exp\left(i\frac{2\pi}{T}tk\right) = \sum_{k=-\infty}^{\infty} \widehat{h}(k)\widehat{f}(k)\exp\left(i\frac{2\pi}{T}tk\right).$$

Therefore, the convolution can be computed by componentwise multiplication in frequency space.

The implementation of the Fourier transform and of the inverse Fourier transform can be done using the Discrete Fourier Transform (DFT) and its fast and memory efficient implementation, the Fast Fourier Transform. As an example, we apply the FFT-based convolution to convolving a box-car function with a Gaussian convolution kernel h,

$$h(t) = \frac{1}{\sqrt{2\pi\gamma^2}}\exp\left(-\frac{1}{2\gamma^2}t^2\right),$$

whose Fourier transform can be calculated analytically,

$$\widehat{h}(k) = \frac{1}{\sqrt{2\pi\gamma^2}} \int_{-\infty}^{\infty} \exp\left(-\frac{1}{2\gamma^2}t^2\right) \exp\left(-i\frac{2\pi}{T}tk\right) dt$$

$$= \exp\left(-\frac{1}{2}\left(\frac{2\pi\gamma}{T}k\right)^2\right).$$

We leave as an exercise (see the end of this chapter) to verify that the convolution can be implemented in Matlab as follows:

```
% Input signal
n = 64;
T = 3;
t = linspace(0,T,n)';
x = (t>1).*(t<2);

% Fourier transform of the Gaussian kernel
xi = (2*pi*gamma)/T*[[0:n/2],[-n/2+1:-1]]';
h = exp(-0.5*xi.^2);

% Convolution h*x
hx = real(ifft(h.*fft(x)));
```

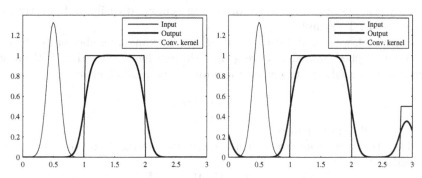

Fig. 4.2. Convolution with a Gaussian kernel computed with the FFT algorithm. The input signal in the example shown in the right panel is not vanishing at the right edge, hence its periodic extension causes an artefact at the left edge.

A couple of comments concerning the code above are in order. First, note that the components of the vector **xi** are ordered according to how the Fourier coefficients are returned by the Matlab function **FFT**; see the exercise at the end of this chapter. Second, notice that when calculating the final result, the Matlab function **real** is used to eliminate imaginary components of the output that are due to numerical round-off.

A second comment concerns the periodic extension of the input signal. In Figure 4.2, we have plotted the output signals corresponding to an input that vanishes near the end points of the interval and to one that does not vanish. In the latter example, the non-locality of the convolution operator causes an artifact at the left end of the output signal. To avoid this artifact, the input signal must be first padded with zero outside the interval $[0, T]$, and the FFT convolution must be then applied to the extended signal. This procedure prevents the convolution kernel from reading a signal that comes from the preceding or following period of f_T. Further properties of the FFT, concerning, for example, the ordering of the components, can be found in the exercises at the end of this chapter.

The philosophy behind solving a linear system via an iterative method is quite different from the one behind the use of direct solvers. Instead of the all-or-nothing approach of direct methods, iterative methods, starting from an initial guess x_0, compute a sequence

$$x_1, x_2, \ldots, x_k \ldots$$

of improved approximate solutions. The matrix A does not need to be explicitly formed, but we must be able to compute its product and possibly the product of its transpose with an arbitrary vector. This flexibility with respect to the matrix A makes iterative linear solvers the methods of choice when the matrix A is either not explicitly known, or its storage would require a lot of memory, but its action on a vector can be computed easily and effectively, like in the FFT example above.

In general, we stop iterating when either a maximum number of iterations has been reached, or some convergence criterion has been satisfied. Usually iterative linear solvers stop when the norm of the residual error

$$r_k = b - Ax_k$$

has been sufficiently reduced.

The kth *Krylov subspace* associated with the matrix A and the vector b is defined to be

$$\mathcal{K}_k(A, b) = \text{span}\{b, Ab, \ldots, A^{k-1}b\}.$$

Iterative methods which seek the kth approximate solution in a Krylov subspace of order k are called *Krylov subspace iterative methods*.

One of the first Krylov subspace methods to be introduced, in 1952, was the *Conjugate Gradient (CG) method*. This iterative method in its original formulation can be applied only to solve linear systems with a symmetric positive definite matrix, although some variants for singular matrices have also been proposed. The popularity of the CG method is due to the fact that it requires only one matrix-vector product per iteration, and the memory allocation is essentially independent of the number of iterations. The kth iterate computed by the CG method minimizes the A norm of the error,

$$x_k = \arg \min_{x \in \mathcal{K}_k(A,b)} \|x - x_*\|_A^2,$$

where x_* stands for the presumably existing but, evidently, *unknown* exact solution of the system, and the A–norm is defined by

$$\|z\|_A^2 = z^{\mathrm{T}} A z.$$

Fortunately, even though we do not know x_* and hence cannot evaluate the functional to be minimized, an algorithm for finding the minimizer x_k is available. A detailed derivation of the algorithm is not presented here, but instead we give an outline of the idea behind it.

At each step, the CG algorithm searches for the scalar α which minimizes the function

$$\alpha \mapsto \|x_{k-1} + \alpha p_{k-1} - x_*\|_A^2,$$

where x_{k-1} is the approximate solution from the previous iteration and p_{k-1} is a search direction. The minimizer turns out to be

$$\alpha_k = \frac{\|r_{k-1}\|^2}{p_{k-1}^{\mathrm{T}} A p_{k-1}},$$

and the new approximate solution is obtained by updating the previous one,

$$x_k = x_{k-1} + \alpha_{k-1} p_{k-1}.$$

The selection of the search directions is crucial. At the first iteration, we search along

$$p_0 = r_0 = b - A x_0,$$

the residual associated with the initial guess. At each subsequent iteration, the new search direction is A–*conjugate* to the previous ones, that is, p_k must satisfy

$$p_k^{\mathrm{T}} A p_j = 0, \quad 0 \le j \le k - 1.$$

At first sight, this requirement looks quite complicated and its numerical implementation time-consuming. It turns out, however, that the new search direction can always be chosen in the subspace spanned by the previous direction and the most recent residual vector, that is,

$$p_k = r_k + \beta_k p_{k-1}, \quad r_k = b - A x_k,$$

where the parameter β_k is chosen so that the A–conjugacy is satisfied. After some algebraic manipulation it turns out that

$$\beta_k = \frac{\|r_k\|^2}{\|r_{k-1}\|^2},$$

We are ready to outline how to organize the computations for the Conjugate Gradient method. For simplicity, the initial guess is set to zero here.

The CG Algorithm: Given the right hand side b, initialize:

$x_0 = 0$;

$r_0 = b - Ax_0$;

$p_0 = r_0$;

Iterative algorithm:

for $k = 1, 2, \ldots$ until the stopping criterion is satisfied

$$\alpha = \frac{\|r_{k-1}\|^2}{(p_{k-1}^T A p_{k-1})};$$

$$x_k = x_{k-1} + \alpha p_{k-1};$$
$$r_k = r_{k-1} - \alpha A p_{k-1};$$

$$\beta = \frac{\|r_k\|^2}{\|r_{k-1}\|^2};$$

$$p_k = r_k + \beta p_{k-1};$$

end

If the matrix A is not symmetric positive definite the Conjugate Gradient method will break down. An algorithm which can be applied to the more general case where the matrix A is not necessarily even square is the *Conjugate Gradient method for Least Squares* (CGLS). This iterative method is mathematically equivalent to applying the CG method to normal equations

$$A^T A x = A^T b,$$

without ever forming the matrix $A^T A$. The CGLS method is computationally more expensive than the CG method, requiring two matrix-vector products per iteration, one with A, one with A^T, but its memory allocation is essentially independent of the number of iterations. It can be shown that the kth CGLS iterate solves the minimization problem

$$x_k = \arg \min_{x \in \mathcal{K}_k(A^T A, A^T b)} \|b - Ax\|^2,$$

or, equivalently, is characterized by

$$\Phi(x_k) = \min_{x \in \mathcal{K}_k(A^T A, A^T b)} \Phi(x),$$

where

$$\Phi(x) = \frac{1}{2} x^T A^T A x - x^T A^T b.$$

The minimization strategy is very similar to that of the Conjugate Gradient method. The search for the minimizer is done by performing sequential linear searches along the $A^T A$–*conjugate directions*,

$$p_0, p_1, \ldots, p_{k-1},$$

where $A^T A$–conjugate means that the vectors p_j satisfy the condition

$$p_j^T A^T A p_k = 0, \quad j \neq k.$$

The iterate x_k is determined from x_{k-1} and p_{k-1} by the formula

$$x_k = x_{k-1} + \alpha_{k-1} p_{k-1},$$

where the coefficient $\alpha_{k-1} \in \mathbb{R}$ solves the minimization problem

$$\alpha_{k-1} = \arg\min_{\alpha \in \mathbb{R}} \Phi(x_{k-1} + \alpha p_{k-1}).$$

Introducing the *residual error of the normal equations* for the CGLS method associated with x_k,

$$r_k = A^T b - A^T A x_k,$$

the passage from the current search direction to the next is given by

$$p_k = r_k + \beta_k p_{k-1},$$

where we choose the coefficient β_k so that p_k is $A^T A$-conjugate to the previous search directions,

$$p_k^T A^T A p_j = 0, \quad 1 \leq j \leq k - 1.$$

It can be shown that

$$\beta_k = \frac{\|r_k\|^2}{\|r_{k-1}\|^2}.$$

The quantity

$$d_k = b - A x_k$$

is called the *discrepancy* associated with x_k. The discrepancy and the residual of the normal equations are related to each other via the equation

$$r_k = A^T d_k.$$

Note that the norms of the discrepancies form a non-increasing sequence, while the norms of the solutions form a non-increasing one,

$$\|d_{k+1}\| \leq \|d_k\|, \qquad \|x_{k+1}\| \geq \|x_k\|.$$

We are now ready to describe how the calculations for the CGLS method should be organized. Here, again, the initial guess is set to zero.

The CGLS Algorithm: Given the right hand side b, initialize:
$x_0 = 0$;
$d_0 = b - A x_0$;
$r_0 = A^T d_0$;
$p_0 = r_0$;

$$y_0 = Ap_0;$$

Iteration:
for $k = 1, 2, \ldots$ until the stopping criterion is satisfied

$$\alpha = \frac{\|r_{k-1}\|^2}{\|y_{k-1}\|^2}$$

$$x_k = x_{k-1} + \alpha p_{k-1};$$
$$d_k = d_{k-1} - \alpha y_{k-1};$$
$$r_k = A^{\mathrm{T}} d_k;$$

$$\beta = \frac{\|r_k\|^2}{\|r_{k-1}\|^2};$$

$$p_k = r_k + \beta p_{k-1};$$
$$y_k = A p_k;$$
end

A more recent addition to the family of Krylov subspace iterative methods, the *Generalized Minimal RESidual* (GMRES) method, can be applied to any linear system provided that the coefficient matrix A is invertible. The computational cost per iteration of the GMRES method is low, only requiring one matrix-vector product, but its memory allocation grows by one vector of the size of x at each iteration.

The kth iterate of the GMRES method solves the minimization problem

$$x_k = \arg \min_{x \in \mathcal{K}_k(A, b)} \|b - Ax\|,$$

a property which gives the method its name. Since for the GMRES method the residual vector is also the discrepancy, from this characterization of the iterates of the GMRES it follows immediately that

$$\|d_k\| \leq \|d_{k-1}\| \leq \ldots \leq \|d_0\| = \|b\|,$$

because the Krylov subspaces where the minimization problems are solved as the iterations progress form a nested sequence. To understand how the GMRES computations are organized, we begin with an algorithm for computing an orthonormal basis of $\mathcal{K}_k(A, b)$ directly. This is accomplished by the following algorithm, called the *Arnoldi process*.

Consider first the Krylov subspace $\mathcal{K}_1(A, b) = \{b\}$. Clearly, an orthonormal basis for this subspace consists of one single vector v_1,

$$v_1 = \frac{1}{\|b\|} b, \quad b \neq 0.$$

Observe that if $b = 0$, $x = 0$ is the solution and no iterations are needed.

Assume now that we have found an orthonormal basis for $\mathcal{K}_k(A, b)$,

$$\{v_1, v_2, \ldots, v_k\}.$$

To augment this set with one additional vector, orthogonal to the previous ones and spanning the extra direction of $\mathcal{K}_{k+1}(A, b)$, introduce the vector

$$w = Av_k,$$

and search for a new vector v_{k+1} of the form

$$v_{k+1} = w - \sum_{j=1}^{k} h_{jk} v_j.$$

We impose that the new vector is orthogonal to the previous ones, by requiring that it satisfies the condition

$$0 = v_j^{\mathrm{T}} v_{k+1} = v_j^{\mathrm{T}} w - h_{jk}, \quad 1 \le j \le k,$$

which, in turn, defines the weights h_{jk}. Finally we scale the vector v_{k+1} to have norm one. Algorithmically, we write the Arnoldi process as follows:

Arnoldi Process:

$v_1 = \dfrac{b}{\|b\|}$;

for $j = 1, \ldots, k$

$\quad w = Av_j$;

$\quad$ for $i = 1, 2, \ldots, j$

$\quad\quad h_{ij} = w^{\mathrm{T}} v_i$;

$\quad\quad w = w - h_{ij} v_j$;

$\quad$ end

$\quad h_{j+1,j} = \|w\|$;

$\quad v_{j+1} = \dfrac{w}{\|w\|}$;

end

Once the Arnoldi process has been completed, the vectors v_j form an orthonormal basis of the kth Krylov subspace, that is

$$v_i^{\mathrm{T}} v_j = \begin{cases} 1, \, i = j \\ 0, \, i \ne j \end{cases}$$

and

$$\mathrm{span}\{v_1, v_2, \ldots, v_{k+1}\} = \mathcal{K}_{k+1}(A, b).$$

We collect the vectors v_i in the matrix

$$V_j = [v_1 \; v_2 \; \cdots \; v_j], j = k, k+1,$$

and the weights h_{ij} in the matrix H,

$$H_k = \begin{bmatrix} h_{11} & h_{12} & & h_{1k} \\ h_{21} & h_{22} & & h_{2k} \\ & \ddots & \ddots & \\ & & & h_{kk} \\ & & & h_{k+1,k} \end{bmatrix}.$$

A matrix of this form is referred to as a Hessenberg matrix.

It follows from the Arnoldi process that these matrices satisfy the *Arnoldi relation*,

$$AV_k = V_{k+1}H_k.$$

Once an orthonormal basis for $\mathcal{K}_k(A,b)$ is available, we can proceed to solve the minimization problem. Using the Arnoldi relation and the fact that the vectors v_j are orthonormal, we deduce that

$$\min_{x \in \mathcal{K}_k(A,b)} \|Ax - b\| = \min_{y \in \mathbb{R}^k} \|AV_k y - b\|$$
$$= \min_{y \in \mathbb{R}^k} \|V_{k+1}(H_k y - \|b\|e_1)\|$$
$$= \min_{y \in \mathbb{R}^k} \|H_k y - \|b\|e_1\|.$$

If y_k is the least squares solution of the system

$$H_k y = \|b\|e_1,$$

which, in turn, can be written in terms of the pseudoinverse $(H_k)^{\dagger}$ of H_k,

$$y_k = \|b\|(H_k)^{\dagger}e_1,$$

then the solution x_k to the original minimization problem is

$$x_k = V_k y_k.$$

We now present an outline of how the computations for the GMRES method should be organized. In this version, the Arnoldi process has been built into the algorithm.

The GMRES Algorithm: Initialize:
$x_0 = 0;$
$r_0 = b - Ax_0;$
$H = [\,];$ (empty matrix)
$V = [\,];$
$v_1 = \dfrac{r_0}{\|r_0\|};$
$j = 0;$

Iteration: for $k = 1, 2, \ldots$ until the stopping criterion is satisfied

    $j = j + 1;$ Add one column to V:

    $V = [V, v_j];$

    $w_j = A v_j;$

    Add one row and one column to H:

    for $i = 1, \ldots, j$

        $h_{ij} = w_j^{\mathrm{T}} v_i;$

        $w_j = w_j - h_{ij} v_i;$

    end

    $h_{j+1,j} = \|w_j\|;$

    if $h_{j+1,j} = 0$

        No independent directions found; stop iteration

    else

$$v_{j+1} = \frac{w_j}{h_{j+1,1}};$$

    end

    $y = \|b\| H^\dagger e_1;$

    $x_j = V y;$

end

The three iterative methods described can be found as built-in function in Matlab. Each one of their calling sequences allows us to either pass the matrix A as an argument, or to specify a function which computes the product of an arbitrary vector with the matrix A or with its transpose. The choice of the maximum number of steps is an integral part of the algorithm, as is the choice of the tolerance value which decides when the approximate solution is sufficiently accurate. If these fields are not specified, default values are used. All these iterative methods allow the use of *preconditioning*, a practice whose aim is usually to improve the convergence rate of the method. Since preconditioners will be used to import the statistical perspective into the calculations, we will discuss them at length in Chapter 6.

4.4 Ill-conditioning and errors in the data

The solution of linear systems of equations coming from real life applications, with real data as a right-hand side vector, often aims at determining the causes of observed effects. Depending on the type of linear relation between the cause x and the effect b expressed by the linear system, and on the discretization used to represent the problem in terms of matrices and vectors, the matrix A in (4.1) may become moderately to severely *ill-conditioned*. Ill-conditioning is best understood in terms of the singular value decomposition of A,

$$A = UDV^{\mathrm{T}}.$$

Consider the least squares solution of the problem (4.1) defined via the pseudoinverse,

$$x_{LS} = \frac{1}{d_1} V \operatorname{diag}\left(1, \frac{d_1}{d_2}, \ldots, \frac{d_1}{d_r}, 0, \ldots, 0\right) U^T b.$$

If b is contaminated by noise so is $U^T b$, whose kth component is multiplied by the ratio d_1/d_k. As this ratio increases with k, the solution of the linear system may be so dominated by amplified error components to become meaningless. We say that the matrix A is ill-conditioned if its condition number

$$\operatorname{cond}(A) = \frac{d_1}{d_r}$$

is very large. What very large means in practice depends on the application. In general, condition numbers of the order of $10^5 \cdots 10^6$ start becoming of concern.

The effect of having singular values of different orders of magnitude is illustrated by a simple example.

EXAMPLE 4.5: In this elementary two-dimensional model, the matrix A is defined via its eigenvalue decomposition,

$$A = \begin{bmatrix} v_1 & v_2 \end{bmatrix} \begin{bmatrix} 1 & \\ & 0.1 \end{bmatrix} \begin{bmatrix} v_1^T \\ v_2^T \end{bmatrix},$$

where the eigenvectors v_1 and v_2 are

$$v_1 = \begin{bmatrix} \cos\theta \\ \sin\theta \end{bmatrix}, \quad v_2 = \begin{bmatrix} \sin\theta \\ -\cos\theta \end{bmatrix}, \quad \theta = \frac{\pi}{6}.$$

Observe that the condition number of A is only 10, so the example should be understood merely as an illustration.

In Figure 4.3, we have plotted the eigenvectors of a matrix A, scaled by the corresponding eigenvalues and applied at a point x_*. The point x_* is then mapped to the point $y_* = Ax_*$, shown in the right panel of the figure. We may think of y_* as the errorless data and x_* as the true solution of the equation

$$Ax = y_*.$$

What happens to the solution if we perturb the errorless solution y_* by a small amount? Consider two different perturbed data points y_1 and y_2, plotted in the right panel of Figure 4.3. The preimages x_j that satisfy

$$Ax_j = y_j, \quad j = 1, 2,$$

are shown in the left panel of Figure 4.3. More generally, we have also plotted a shaded ellipse, which is the preimage of the disc on the left. Although the distance of y_1 and y_2 from y_* is the same, the distance of the preimages from x_* are very different. In fact, in the eigendirection corresponding to the small eigenvalue, the sensitivity to data errors is tenfold compared to that in the direction corresponding to the larger eigenvalue.

The solution of ill-conditioned linear systems by direct methods is usually not advisable, since the "all-or-nothing" approach is not able to separate the meaningful part of the computed solution from the amplified error components. The design of algorithms for the solution of linear systems with singular values spanning a broad range and clustering around the origin, often referred to as linear discrete ill-posed problems, is an active area of research in inverse problems.

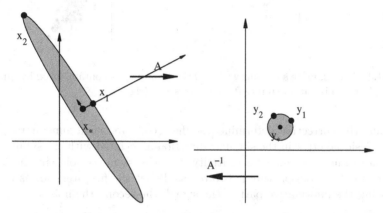

Fig. 4.3. Ill-conditioning and inversion. The noisy data points y_1 and y_2 are equally close to the noiseless vector $y_* = Ax_*$, but this is not the case for the distance of their preimages $x_1 = A^{-1}y_1$ and $x_2 = A^{-1}y_2$ from x_* .

When using iterative methods for the solution of this class of linear systems, a *semiconvergence* behavior can often be observed. At first the iterates seems to converge to a meaningful solution, but as the iterations proceed, they begin to diverge. This phenomenon is illustrated first using the above two-dimensional toy problem.

EXAMPLE 4.6: Starting from the two-dimensional linear system of Example 4.5, consider the additive noise model

$$B = Ax + E, \quad E \sim \mathcal{N}(0, \sigma^2 I),$$

and generate a sample of data vectors, $b_1, b_2, \ldots, b_n$. Each linear system $Ax = b_i$ with right hand side a vector in the sample is then solved with the Conjugate Gradient method. As expected, only two iteration steps are needed. In Figure 4.4, we show the approximate solutions obtained after one (left panel) and two (right panel) iteration steps.

The results reveal some typical features of iterative solvers. In fact, the correction step determined in the first iteration aims at reducing the residual error in the singular direction associated with the larger singular value,

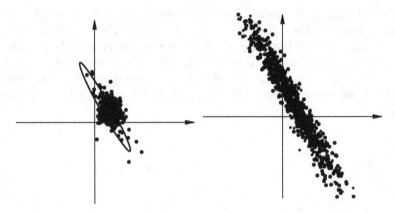

Fig. 4.4. Approximate solutions with 100 right hand sides contaminated by random error, after one iteration step (left) and two steps (right).

while the correction determined in the second iteration aims at reducing the residual error in the singular direction associated with the smaller singular value. The higher sensitivity of the system to noise in the latter direction is a reason why it is often said that the first iteration is mostly fitting the informative part of the signal, the second the noise.

This last example, in spite of its simplicity, shows how the idea of capturing the solution before the amplified noise takes over by truncating the iteration prior to reaching convergence came about. By equipping an iterative method with a stopping rule effective at filtering the amplified noise from the computed solution, we can make the problem less sensitive to perturbations in the data. The process of stabilizing the solution of a discrete ill-posed problem is usually referred to as *regularization*. The process of regularizing by prematurely stopping the iteration process is called *regularization by truncated iteration*.

The point where the iterations start to fit the noise, or when the convergence switches to divergence, varies from one iterative method to the other, but in general it seems to occur sooner for the GMRES method.

The design of a good stopping rule is an important and difficult task. We outline here how this problem can be approached in a fully deterministic way, prior to revisiting the issue from a Bayesian perspective.

Assume for the moment that we are considering a linear system with an invertible square matrix A, and that all our information about the noise in the right hand side amounts to an estimate of its norm. Denoting by x_* the exact solution of the system with noiseless data b_*, we may write

$$Ax = b_* + \varepsilon = Ax_* + \varepsilon = b,$$

where ε represents the inaccuracy of the data, and we have the approximate information that

$$\|\varepsilon\| \approx \eta,$$

with $\eta > 0$ known. The fact that

$$\|A(x - x_*)\| = \|\varepsilon\| \approx \eta$$

seems to suggest that, within the limits of our information, any solution x that satisfies the condition

$$\|Ax - b\| \leq \tau\eta \tag{4.6}$$

should be a good approximate solution. Here the *fudge factor* $\tau > 1$, whose value is close to one, safeguards against underestimating the norm of the noise. The idea of using (4.6) to decide how much a problem should be regularized is called the *Morozov discrepancy principle*. We now present a few examples illustrating the use of this principle in classical deterministic regularization.

EXAMPLE 4.7: In this example we consider regularization by truncated iteration with iterative solvers applied to a linear system

$$b = Ax + e,$$

which is the discretization of a deconvolution problem. The matrix A is of Toeplitz form and it is the discretization of a Gaussian kernel. The following Matlab code can be used to generate the matrix:

```
n = 64;                 %  Number of discretization points
T = 3;                  %  Length of the interval
t = linspace(0,T,n)';
gamma = 0.2;            %  Width parameter
hh = (T/n)*1/(sqrt(2*pi*gamma^2))*
        exp(-1/(2*gamma^2)*t.^2);
A = toeplitz(hh,hh);
```

Since this matrix A is extremely ill-conditioned, with condition number

$$\text{cond}(A) \approx 5.7 \times 10^{18},$$

we expect direct solvers without regularization to be completely useless for the solution of this problem. To verify this, we take as true input signal x_* the boxcar function of Example 4.4, calculate the errorless data, $b = Ax_*$, and solve the equation $b = Ax$ by a direct solver. The result, shown in Figure 4.5, is indeed meaningless even if the only source of error was just numerical rounding of the data.

We investigate next how iterative solvers perform. In fact, confident in their performance, we add Gaussian random error with standard deviation $\sigma = 10^{-4}$ to the errorless data. We start with the CG method, which can be applied to this problem because we know that the matrix A is

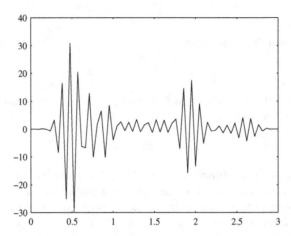

Fig. 4.5. The solution of the deconvolution problem by direct linear system solver using errorless data corresponding to the boxcar input.

mathematically positive definite. To illustrate the semiconvergence phenomenon, in Figure 4.6 we plot the norm of the solution error and the norm of the discrepancy against the iteration index

$$k \mapsto \|x_k - x_*\|, \quad k \mapsto \|r_k\| = \|b - Ax_k\|, \quad k = 0, 1, 2, \ldots$$

In real applications only the norm of the residual error can be calculated, because the error of the approximation is not available unless we know x_*, the true solution[1]. The sequence of approximate solutions computed by the CG iterations are shown in Figure 4.6.

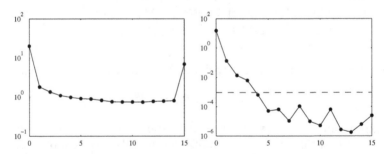

Fig. 4.6. The norm of the error (left) and of the discrepancy (right) of the Conjugate Gradient iterates as functions of the iteration index. The dashed line indicates the norm of the noise in the right hand side.

From the plot of the error and residual norms it is clear that the norm of the residual error associated with the fourth iterate x_4 is at the level

---

[1] and, to state the obvious, if we knew x_* we would not need to estimate it!

of the norm of the noise. If we apply the Morozov discrepancy principle, we stop around the fourth or fifth iteration. The semiconvergence is quite noticeable in this example: at the beginning, the error decreases but after the thirteenth iteration, the noise starts to take over and the error starts to grow. To see how the iterations proceed, we plot the fifteen first iterations in Figure 4.7.

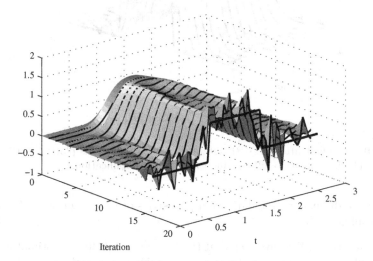

Fig. 4.7. The first fifteen Conjugate Gradient iterates. The true input (boxcar function) is plotted in the foreground. Notice how the last iterate oscillates due to noise amplification.

For comparison, we repeat the procedure using the GMRES iteration. The results are displayed in Figures 4.8 and 4.9.

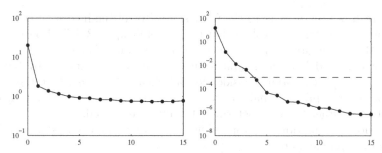

Fig. 4.8. The error (left) and the discrepancy (right) norm of the GMRES iteration as functions of the iteration index. The dashed line indicates the norm of the noise in the right hand side.

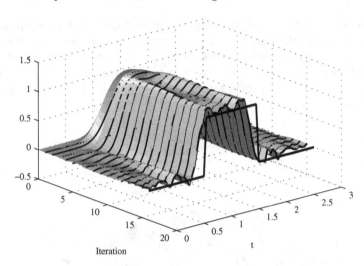

Fig. 4.9. The first fifteen GMRES iterates. The true input is plotted in the foreground. The last iterate oscillates due to the noise amplification.

Now we discuss the regularization properties of using a prior in the inference problem.

EXAMPLE 4.8: We have seen that the Maximum Likelihood estimator with a single noisy observation leads to the minimization problem

$$x_{\mathrm{LS}} = \arg\min \|Ax - b\|,$$

while the Maximum A Posteriori estimator for this linear model with a Gaussian prior, leads to

$$x_{\mathrm{MAP}} = \arg\min \left\| \begin{bmatrix} A \\ \delta R \end{bmatrix} x - \begin{bmatrix} b \\ 0 \end{bmatrix} \right\|,$$

where R is proportional to the Cholesky factor of the inverse of the prior covariance matrix, as explained at the end of Chapter 3. We also saw that the MAP estimate is the minimizer of the Tikhonov functional.
To understand how the regularization properties are affected by the introduction of the prior, consider for simplicity the case where the prior covariance is proportional to the unit matrix and $m \geq n$. That is, the system is either formally determined or overdetermined.
The normal equations corresponding to the Maximum A Posteriori minimization problem are

$$\begin{bmatrix} A \\ \delta I \end{bmatrix}^{\mathrm{T}} \begin{bmatrix} A \\ \delta I \end{bmatrix} x = \begin{bmatrix} A \\ \delta I \end{bmatrix}^{\mathrm{T}} \begin{bmatrix} b \\ 0 \end{bmatrix},$$

or, equivalently,

$$(A^{\mathrm{T}}A + \delta^2 I)x = A^{\mathrm{T}}b.$$

To study the ill-conditioning of this system, we replace A with its singular value decomposition

$$A = UDV^{\mathrm{T}},$$

in the left hand side of the normal equations to obtain

$$A^{\mathrm{T}}A + \delta^2 I = VD^{\mathrm{T}}DV^{\mathrm{T}} + \delta^2 I$$

$$= V(D^{\mathrm{T}}D + \delta^2 I)V^{\mathrm{T}}.$$

We remark that

$$D^{\mathrm{T}}D + \delta^2 I = \mathrm{diag}(d_1^2 + \delta^2, \ldots, d_n^2 + \delta^2),$$

hence the prior adds a positive multiple of the identity to the matrix to reduce the ill-conditioning of the problem. In fact, we can write the Maximum A Posteriori solution in terms of the singular values as

$$x_{\mathrm{MAP}} = \sum_{j=1}^{n} \frac{d_j}{d_j^2 + \delta^2}(U^{\mathrm{T}}b)_j v_j,$$

where v_j is the jth column of the matrix V.

A large portion of the classical regularization literature is concerned with the selection of the regularization parameter δ in the Tikhonov functional. One of the most common ways of determining it is to use the Morozov discrepancy principle. Writing the solution of the normal equations as

$$x_\delta = (A^{\mathrm{T}}A + \delta^2 I)^{-1}A^{\mathrm{T}}b,$$

the discrepancy principle suggests that we should choose the largest $\delta > 0$ for which

$$\|Ax_\delta - b\| \leq \tau\eta$$

holds. Of course, if the dimensions of the linear system are large, it may be unfeasible to compute x_δ repeatedly, and more efficient ways of estimating the value of the parameter have to be developed. We do not discuss those issues here.

As a concrete application, consider the problem of numerical differentiation. Let $f : [0,1] \to \mathbb{R}$ be a piecewise smooth function, $f(0) = 0$, and $f'(t)$ be its derivative where defined. Assume that we have noisy observations of f at a few discrete point, that is

$$y_j = f(t_j) + e_j, \quad t_j = \frac{1}{n}j, \quad 1 \leq j \leq n,$$

where e_j denotes the noise. We want to estimate the derivative f' at the n discrete points t_j. To set up our model, we write a piecewise constant approximation of the derivative,

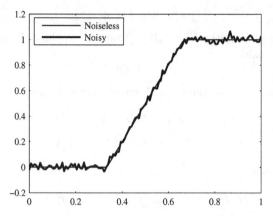

Fig. 4.10. The noiseless data (thin) and data corrupted by noise with standard deviation $\sigma = 0.02$ (thick).

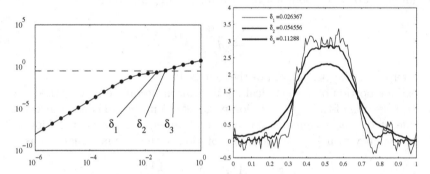

Fig. 4.11. The norm of the discrepancy with different values of the regularization parameter δ (left). The horizontal dashed line corresponds to the discrepancy level $\tau\eta$, where $\tau = 1.5$. The right panel displays reconstructions with three different values of δ, indicated in the left panel in different gray levels. The true signal is a boxcar function of height 3 on the interval $[1/3, 2/3]$.

$$f'(t) = x_j \quad \text{for } t_{j-1} < t \leq t_j,\ 1 \leq j \leq n,$$

where $t_0 = 0$. From the approximation

$$f(t_j) = \int_0^{t_j} f'(t)dt \approx \frac{1}{n}\sum_{k=1}^{j} x_k,$$

we derive the discrete model

$$y = Ax + e,$$

where the matrix A is the lower triangular matrix

$$A = \frac{1}{n} \begin{bmatrix} 1 & & & \\ 1 & 1 & & \\ \vdots & & \ddots & \\ 1 & 1 & \cdots & 1 \end{bmatrix}.$$

The Matlab code for generating the matrix A is

```
col = ones(n,1);
row = zeros(1,n); row(1) = 1;
A = (1/n)*toeplitz(col,row);
```

We calculate A of size 128×128. The condition number of such matrix is not particularly bad, only a few hundreds, but if the noise is significant, the solution without regularization is quite useless.

We generate a piecewise linear noiseless signal which corresponds to the integral of a boxcar function, and add noise to it, see Figure 4.10. Then we solve the regularized problem with different values of δ, ranging in the interval $10^{-6} \cdots 1$. In Figure 4.11, we have plotted the norm of the discrepancy

$$d_k = \|Ax_{\delta_k} - b\|$$

for a few values of $\delta = \delta_k$. We select three of these values, close to $\tau\eta$, where η is the norm of the particular realization of the noise vector and $\tau = 1.5$. In the same figure we also plot the corresponding estimates of the derivatives. We see that above the noise limit, the solution starts to lose details – an effect called *overregularization* in the literature – while below the noise is strongly present – an effect called *underregularization*. The discrepancy principle seems to give a reasonable solution in this case.

In the deterministic setting, when the norm of the noise is not known, some of the stopping criteria proposed have been based on the Generalized Cross Validation. Another strategy is to detect of amplified noise components in the computed solution by monitoring the growth of its norm via the so called L-curve. We will not discuss these methods here.

As a final comment, consider the normal equations corresponding to an ill-posed system. Let $A \in \mathbb{R}^{m \times n}$ be a matrix that has the singular value decomposition

$$A = UDV^{\mathrm{T}}, \quad D = \mathrm{diag}(d_1, d_2, \ldots, d_{\min(n,m)}),$$

implying that the matrix $A^{\mathrm{T}}A$ appearing in the normal equations admits the decomposition

$$A^{\mathrm{T}}A = VD^{\mathrm{T}} \underbrace{U^{\mathrm{T}}U}_{=I} DV^{\mathrm{T}} = VD^{\mathrm{T}}DV^{\mathrm{T}}.$$

Here, the matrix $D^{\mathrm{T}}D \in \mathbb{R}^{n \times n}$ is diagonal, and the diagonal entries which are the eigenvalues of $A^{\mathrm{T}}A$ are $d_1^2, \ldots, d_{\min(n,m)}^2$, augmented with trailing zeros if $m < n$. From this analysis, it is obvious that the condition number of the matrix may be dramatically larger than that of A, which leads to the conclusion that forming the matrix $A^T A$ may not be always advisable.

Exercises

1. If f is a sample from a T-periodic continuous signal, we approximate numerically its Fourier coefficients by writing

$$\widehat{f}(k) = \frac{1}{T} \int_0^T \exp\left(-i\frac{2\pi}{T}tk\right) f(t)dt$$

$$\approx \frac{1}{T} \frac{T}{n} \sum_{j=0}^{n-1} \exp\left(-i\frac{2\pi}{T}t_j k\right) f(t_j)$$

$$= \frac{1}{n} \sum_{j=0}^{n-1} W_n^{-jk} x_j, \quad W_n = \exp\left(i\frac{2\pi}{n}\right).$$

The transformation

$$\mathbb{C}^n \to \mathbb{C}^n, \quad x \mapsto \left[\frac{1}{n} \sum_{j=0}^{n-1} W_n^{-jk} x_j\right]_{1\leq k\leq n},$$

is called the *Discrete Fourier Transform*, and we denote it by $DFT(x)$. This linear operation allows a fast numerical implementation that is based on the structure of the matrix multiplying x. The implementation is called *Fast Fourier Transform* (FFT).

Show that if $x_j^{(\ell)}$ is a discretization of the single frequency signal

$$f^{(\ell)}(t) = \exp\left(i\frac{2\pi}{T}\ell t\right), \quad \ell = 0, \pm 1, \pm 2, \ldots,$$

we get

$$DFT(x^{(\ell)})_k = \begin{cases} 1, & k = \ell \pmod{n}, \\ 0 & \text{otherwise.} \end{cases}$$

Deduce, in particular, that if f is a *band limited signal* of the form

$$f(t) = \sum_{\ell=-n/2+1}^{n/2} \widehat{f}(\ell) f^{(\ell)}(t),$$

the corresponding discretized signal

$$x = \sum_{\ell=-n/2+1}^{n/2} \widehat{f}(\ell) x^{(\ell)},$$

has discrete Fourier transform

$$DFT(x) = \begin{bmatrix} \widehat{f}(0) \\ \widehat{f}(1) \\ \vdots \\ \widehat{f}(n/2) \\ \widehat{f}(-n/2+1) \\ \vdots \\ \widehat{f}(-1) \end{bmatrix}. \tag{4.7}$$

In other words, the DFT returns the Fourier coefficients *exactly* but permuting the order as indicated above.

2. Based on the result of the previous exercise, show that if the spectrum of a signal is not contained in a band of width n, the spectrum calculated with the DFT with n discretization points is *aliased*, i.e., the Fourier coefficients that are offset by n are added together; see Figure 4.12 for an illustration of this phenomenon.

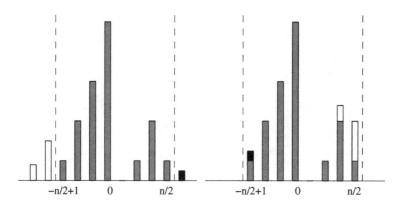

Fig. 4.12. True spectrum of the signal that is sampled with frequency n (left), and the resulting aliased spectrum based on too low sampling frequency (right).

3. The *Inverse Discrete Fourier Transform* and denoted by IDFT, is defined as

$$IDFT(y)_k = \sum_{j=0}^{n-1} W_n^{jk} y_j, \quad 0 \le k \le n-1.$$

Show that

$$IDFT(DFT(x)) = DFT(IDFT(x)) = x.$$

4. Write a Matlab code for implementing the Conjugate Gradient, CGLS and GMRES methods. Include the option that the intermediate iterates x_k can be saved and returned as output, as these are important later to understand the use of iterative solvers as regularization method.

5

Sampling: first encounter

The basic problem of statistical inference is to infer on an unknown distribution based on a sample that is believed to come from that distribution. In this section, we start our journey in the opposite direction. That is, we want to generate a sample from a given distribution. Why is such a task interesting or important? There are at least two reasons.

It has been demonstrated on several occasions in the previous chapters that drawing samples from a given distribution helps us understand and visualize the problems that are being studied. In particular, when we write a prior probability density starting from a prior belief that from the beginning is not quantitative, we need to investigate whether the prior distribution really corresponds to what we believe. A useful way to check the quality of the prior is to draw a random sample and look at the realizations. This could be called a visual inspection of the qualities of the prior.

The second important use of sampling is related to integration. Suppose that we have a probability density π and we are asked to estimate the integral

$$I = \int f(x)\pi(x)dx.$$

Estimating integrals by numerical quadratures is a classical problem: one defines quadrature points x_j and corresponding weights w_j dictated by the quadrature rule, then writes the approximation

$$I \approx \sum_{j=1}^{N} w_j f(x_j).$$

Problems arise if the dimensionality of the variable $x \in \mathbb{R}^n$ is high. For instance, approximating integrals by Gauss quadrature of order k over a rectangle in $\mathbb{R}^n$ containing the support of f requires k discretization points per direction, i.e., in the above formula $N = k^n$. It is quite obvious that for n large, the computational burden required is beyond reasonable. So how does sampling help? Assume that instead of the quadrature points, we draw a sample $S = \{x_1, x_2, \ldots, x_N\}$ of mutually independent realizations $x_j \in \mathbb{R}^n$ that are distributed according to π. Then, the Law of Large Numbers tells us that

$$\lim_{N \to \infty} \underbrace{\frac{1}{N} \left(f(x_1) + f(x_2) + \cdots + f(x_N) \right)}_{=I_N} = I$$

almost certainly, and the Central Limit Theorem states that asymptotically,

$$\text{var}\left(I_N - I \right) \approx \frac{\text{var}(f(X))}{N},$$

that is, *regardless of the dimensionality of the problem*, the error goes to zero like $1/\sqrt{N}$. There is another reason why sampling helps: to apply quadrature rules, one has to find first a reasonable region that contains the essential part of the density π. If n is large, this may be a frustrating task not to mention that finding such region is often already a piece of valuable information and object of the statistical inference.

Integration based on random sampling is called *Monte Carlo integration*. Although it looks very attractive, it has indeed its share of difficulties; otherwise one might be led to believe that the world is too perfect. One of the problems is how to draw the sample. We start studying this problem in this section, and we will return to it later with more sophisticated tools at our disposal.

There are two elementary distributions that we take for granted in the sequel. We assume that we have access to

- a random generator from the standard normal distribution,

$$\pi(x) = \frac{1}{\sqrt{2\pi}} \exp\left(-\frac{1}{2} x^2 \right),$$

- a random generator from the uniform distribution over the interval $[0, 1]$,

$$\pi(x) = \chi_{[0,1]}(x).$$

These random generators in Matlab are called with the commands **randn** and **rand**, respectively.

5.1 Sampling from Gaussian distributions

To generate random draws from a Gaussian density, we transform the density into the standard normal density by a process called *whitening*. If X be a multivariate Gaussian random variable, $X \sim \mathcal{N}(x_0, \Gamma)$, with the probability density

$$\pi(x) = \left(\frac{1}{(2\pi)^n \det(\Gamma)}\right)^{1/2} \exp\left(-\frac{1}{2}(x-x_0)^{\mathrm{T}}\Gamma^{-1}(x-x_0)\right),$$

and

$$\Gamma^{-1} = R^{\mathrm{T}}R$$

is the Cholesky decomposition of the inverse of Γ, the probability density of X can be written as

$$\pi(x) = \left(\frac{1}{(2\pi)^n \det(\Gamma)}\right)^{1/2} \exp\left(-\frac{1}{2}\|R(x-x_0)\|^2\right).$$

Define a new random variable,

$$W = R(X - x_0), \tag{5.1}$$

whose probability density is

$$\begin{aligned}
\mathrm{P}\{W \in B\} &= \mathrm{P}\{R(X-x_0) \in B\} \\
&= \mathrm{P}\{X \in R^{-1}(B) + x_0\} \\
&= \left(\frac{1}{(2\pi)^n \det(\Gamma)}\right)^{1/2} \int_{R^{-1}(B)+x_0} \exp\left(-\frac{1}{2}\|R(x-x_0)\|^2\right) dx.
\end{aligned}$$

The change of variable,

$$w = R(x - x_0), \quad dw = |\det(R)|dx,$$

and the observation that

$$\frac{1}{\det(\Gamma)} = \det(\Gamma^{-1}) = \det(R^{\mathrm{T}})\det(R) = (\det(R))^2,$$

lead to the formula

$$\mathrm{P}\{W \in B\} = \frac{1}{(2\pi)^{n/2}} \int_B \exp\left(-\frac{1}{2}\|w\|^2\right) dw,$$

i.e., W is *Gaussian white noise*[1]

$$W \sim \mathcal{N}(0, I).$$

The transformation (5.1) is therefore called *whitening* of X, and the Cholesky factor R of the inverse of the covariance the *whitening matrix*.

If the whitening matrix is known, random draws from a general Gaussian density can be generated as follows:

---

[1] The name white noise has its roots in signal processing: if we think of X as a signal sequence of length n, its FFT is also a random variable with covariance the identity. The interpretation of this assertion is that all frequencies of X have the same average power, so when we "play" a realization of X, it sounds like noise with high and low frequencies in balance.

1. Draw $w \in \mathbb{R}^n$ from the Gaussian white noise density,
2. Find $x \in \mathbb{R}^n$ by solving the linear system

$$w = R(x - x_0).$$

The process is put into action in following example.

EXAMPLE 5.1: This example not only demonstrates the generation of random draws by whitening but also discusses the construction of *structural prior densities*. It is often the case that we seek to estimate a parameter vector that we believe belongs to a class characterized by a structure, whose exact details may not be fully specified. It is natural to think that, if we know the structure, importing the information via the prior would simplify the task of computing the solution of the problem and improve its quality.

To clarify this notion and to explain how to construct a structural prior, we go back to the bacteria density estimation, discussed in Example 3.8. As previously, assume that the medium where the bacteria live is a square array subdivided into square subunits, but now instead of one, we have two different cultures, which causes the density of bacteria to be different in different regions of the square, see Figure 5.1. We also assume that the areas of the two cultures are not perfectly isolated, and this affects the density of bacteria around the boundary between these regions.

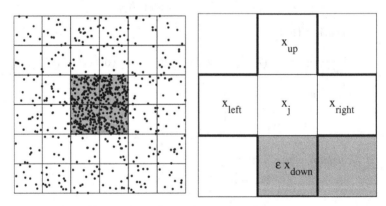

Fig. 5.1. A square array with two different subdomains (left). The coupling constant ε determines the strength of the coupling across the borderline (right).

As before, we assume that the density of bacteria in each square *inside one type of culture* varies smoothly, thus if X_j is the number of bacteria in the jth square, we write a stochastic model

$$X_j = \frac{1}{4}(X_{\text{up}} + X_{\text{down}} + X_{\text{left}} + X_{\text{right}}) + R; \tag{5.2}$$

where R is the innovation term. This formula holds for all interior pixels in both cultures. Consider a *borderline* pixel, whose "up", "right" and "left" neighbors are in the same culture, but whose "down" neighbor is in the other culture, see Figure 5.1. Since the density in the "down" square is presumably different, it should not enter the averaging process in the same way as that of the other neighbors. However, since the isolation between the two cultures is not perfect, it should not be completely left out either. Therefore we perform a *weighted* averaging, where densities from the other culture are given a lower weight,

$$X_j = \frac{1}{3+\varepsilon}(X_{\text{up}} + \varepsilon X_{\text{down}} + X_{\text{left}} + X_{\text{right}}) + R. \qquad (5.3)$$

The factor ε expresses how strongly, or weakly, coupled we believe the two cultures to be; the closer to zero it is the more uncoupled. Similarly, for corner pixels with two neighbors in the other culture, we will use the averaging

$$X_j = \frac{1}{2+2\varepsilon}(X_{\text{up}} + \varepsilon X_{\text{down}} + \varepsilon X_{\text{left}} + X_{\text{right}}) + R. \qquad (5.4)$$

Assuming that the innovation term R in all three cases (5.2)–(5.4) is Gaussian with variance σ^2 and is independent from pixel to pixel as in Example 3.8, we arrive at the Gaussian prior model

$$\pi_{\text{prior}}(x \mid \sigma^2) \propto \exp\left(-\frac{1}{2\sigma^2}\|L_\varepsilon x\|^2\right) = \exp\left(-\frac{1}{2\sigma^2}x^{\text{T}}(L_\varepsilon^{\text{T}} L_\varepsilon)x\right),$$

where

$$L_\varepsilon = I - A_\varepsilon \in \mathbb{R}^{N \times N}$$

follows (5.2) – (5.4).

The inverse of the covariance matrix is immediately available: in fact, we have

$$\Gamma^{-1} = \frac{1}{\sigma^2}L_\varepsilon^{\text{T}} L_\varepsilon.$$

To calculate the whitening matrix, we could compute the Cholesky factor of this matrix. However, since L_ε is invertible, we do not need to compute anything: in fact, the random variable defined as

$$W = \frac{1}{\sigma}L_\varepsilon X$$

is Gaussian white noise!

How does one form the matrices A_ε and L_ε? If we have already calculated the matrix A of Example 3.8, the modification is almost trivial. To see this, consider the situation illustrated in Figure 5.1: the jth pixel has all neighbors but one in the same domain. This means that the jth row of the matrix A is modified as follows:

$$\begin{array}{cccc} \text{(up)} & \text{(down)} & \text{(left)} & \text{(right)} \end{array}$$
$$A(j, :) = \begin{bmatrix} 0 \cdots 1/4 \cdots & 1/4 \cdots & 1/4 \cdots & 1/4 \cdots 0 \end{bmatrix}$$
$$\longrightarrow A_\varepsilon(j, :) = \frac{4}{3+\varepsilon} \begin{bmatrix} 0 \cdots 1/4 \cdots & \varepsilon/4 \cdots & 1/4 \cdots & 1/4 \cdots 0 \end{bmatrix},$$

i.e., the elements that point to neighbors belonging to different domain are multiplied by ε. This leads to a simple algorithm: first, the pixels are reordered so that first come the pixels in the first domain, then the pixels in the second domain. We write a partitioning

$$x = \begin{bmatrix} x^{(1)} \\ x^{(2)} \end{bmatrix},$$

where $x^{(j)}$ is a vector containing the pixel values in domain j, $j = 1, 2$, and we partition the matrix A accordingly,

$$A = \begin{bmatrix} A_{11} & A_{12} \\ A_{21} & A_{22} \end{bmatrix}.$$

The diagonal blocks A_{11} and A_{22} contain the averaging coefficients within the domains, while the off-diagonal ones contain the coefficients across the domain boundaries. Therefore, we have

$$A_\varepsilon = D_\varepsilon \underbrace{\begin{bmatrix} A_{11} & \varepsilon A_{12} \\ \varepsilon A_{21} & A_{22} \end{bmatrix}}_{=\tilde{A}_\varepsilon},$$

where D_ε is a diagonal matrix whose diagonal is the inverse of the row sum of the matrix $\tilde{A}_\varepsilon$.

Below is one possible way of constructing the matrix L_ε.

```
n = 50;   % Number of pixels per direction

% Define the structure: a rectangle R = [k1,k2]x[k1,k2]

k1 = 15; k2 = 35;

% Create an index vector Iin that contains indices
% to squares within the inner domain. The remaining
% indices are in Iout.

I = [1:n^2];
I = reshape(I,n,n);
Iin = I(k1:k2,k1:k2);
Iin = Iin(:);
```

```
Iout = setdiff(I(:),Iin);

% Smoothness prior with Dirichlet boundary without the
% structure: Calculate A as in Example 3.8 (omitted
% here).

% Defining the weak coupling constant epsilon and
% uncoupling

epsilon = 0.1;
A(Iin,Iout) = epsilon*A(Iin,Iout);
A(Iout,Iin) = epsilon*A(Iout,Iin);

% Scaling by a diagonal matrix

d = sum(A')';
A = spdiags(1 ./d,0,speye(n^2))*A;
L = speye(n^2) - A;
```

In order to avoid the unnecessary crowding of the computer memory, we define directly the factor of the inverse of the covariance matrix as a sparse matrix. The role of the index sets Iin and Iout is fundamental, because it defines the spatial location of the structure. In our example, the indices of the pixels corresponding to the inner structure belong to Iin and the indices of the pixels outside this structure are in Iout. Note how we construct Iout from Iin using the logical Matlab command setdiff.

Once the structural whitening matrix has been computed, we can perform random draws from the corresponding prior with different values of ε. Here is a section of Matlab code that does the random draws and plots them as surfaces and as images.

```
ndraws = 4;
for j = 1:ndraws
    w = randn(n*n,1);
    x = L\w;
    figure(j)
    surfl(flipud(reshape(x,n,n)))
    shading interp
    colormap(gray)
    figure(ndraws+j)
    imagesc((reshape(x,n,n))
    axis('square')
    axis('off')
    shading flat
    colormap(0.95*(1-gray))
end
```

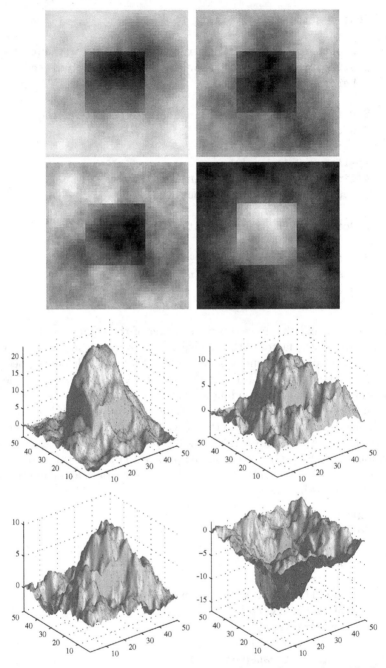

Fig. 5.2. Four random draws from the structural smoothness prior with the coupling parameter $\varepsilon = 0.1$.

Figure 5.2 shows four random draws from a Gaussian density with a structural prior constructed as described above and coupling parameter equal to $\varepsilon = 0.1$.

5.2 Random draws from non-Gaussian densities

We have thus seen that random draws from Gaussian distributions seem to be under control. But what about non-Gaussian ones? As a prelude to the Markov Chain Monte Carlo (MCMC) techniques to be introduced later, let us look at two model algorithms. They are based on the following ideas:

(a) Draw the random sample *directly* from the actual distribution,

(b) Find a fast approximate way of drawing, and correct later the sample in some way.

The former approach is attractive, but it turns out to be costly unless the probability density is one of standard form (e.g. normal) for which effective random generators are available. The success of the second approach depends entirely of how smart *proposal distribution* we are able to design, and might require a lot of hand tuning.

We start with the approach (a), restricting the discussion, for the time being, to one-dimensional densities. Let X be a real valued random variable whose probability density is $\pi(x)$. For simplicity, we assume that in some interval, finite or infinite, $\pi(x) > 0$, and $\pi(x) = 0$ possibly only at isolated points. Define the cumulative distribution function,

$$\Phi(z) = \int_{-\infty}^{z} \pi(x)dx.$$

Clearly, Φ is non-decreasing since $\pi \geq 0$, and $0 \leq \Phi \leq 1$. In fact, the assumption about strict positivity of π implies that Φ is strictly increasing in some interval.

Let us define a new random variable,

$$T = \Phi(X). \tag{5.5}$$

We claim that $T \sim \text{Uniform}([0, 1])$. To see this, observe first that, due to the monotonicity of Φ,

$$P\{T < a\} = P\{\Phi(X) < a\} = P\{X < \Phi^{-1}(a)\}, \quad 0 < a < 1.$$

On the other hand, by the definition of the probability density,

$$P\{X < \Phi^{-1}(a)\} = \int_{-\infty}^{\Phi^{-1}(a)} \pi(x)dx = \int_{-\infty}^{\Phi^{-1}(a)} \Phi'(x)dx.$$

After the change of variable

$$t = \Phi(x), \quad dt = \Phi'(x)dx,$$

we arrive at

$$P\{T < a\} = \int_{-\infty}^{\Phi^{-1}(a)} \Phi'(x)dx = \int_0^a dt = a,$$

which is the very definition of T being a random variable with uniform distribution.

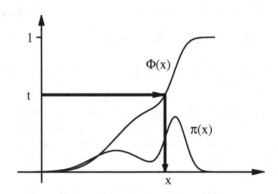

Fig. 5.3. Draw of x from the distribution $\pi(x)$. Here, $t \sim \mathrm{Uniform}([0,1])$.

We have now an algorithm to draw from the distribution π, graphically illustrated in Figure 5.3:

1. Draw $t \sim \mathrm{Uniform}([0,1])$,
2. Calculate $x = \Phi^{-1}(t)$.

This algorithm, sometimes referred to as the *Golden Rule* or *Inverse Cumulative Distribution Rule*, is very useful also in multivariate cases. Attractive and innocent as it looks, its implementation can present some problems, as the following example shows.

EXAMPLE 5.2: Consider a one-dimensional normal distribution with a bound constraint,

$$\pi(x) \propto \pi_c(x)\exp\left(-\frac{1}{2}x^2\right),$$

where

$$\pi_c(x) = \begin{cases} 1, & \text{if } x > c, \\ 0, & \text{if } x \le c, \end{cases} \quad c \in \mathbb{R},$$

from which we want to generate a sample.
The cumulative distribution function is

$$\Phi(z) = C \int_c^z e^{-x^2/2} dx,$$

where $C > 0$ is a normalizing constant,

$$C = \left(\int_c^\infty e^{-x^2/2} dx \right)^{-1}.$$

The function Φ has to be calculated numerically. Fortunately, there are routines readily available to do the calculation: in Matlab, the built-in *error function*, erf, is defined as

$$\mathrm{erf}(t) = \frac{2}{\sqrt{\pi}} \int_0^t e^{-s^2} ds.$$

We observe that

$$\Phi(z) = C \left(\int_0^z - \int_0^c \right) e^{-x^2/2} dx = \sqrt{2} C \left(\int_0^{z/\sqrt{2}} - \int_0^{c/\sqrt{2}} \right) e^{-s^2} ds$$

$$= \sqrt{\frac{\pi}{2}} C \left(\mathrm{erf}(z/\sqrt{2}) - \mathrm{erf}(c/\sqrt{2}) \right).$$

On the other hand, since

$$C = \left(\sqrt{\frac{\pi}{2}} C \left(1 - \mathrm{erf}(c/\sqrt{2}) \right) \right)^{-1},$$

we have

$$\Phi(z) = \frac{\left(\mathrm{erf}(z/\sqrt{2}) - \mathrm{erf}(c/\sqrt{2}) \right)}{\left(1 - \mathrm{erf}(c/\sqrt{2}) \right)}.$$

What about the inverse? Writing

$$\Phi(x) = t,$$

after some algebraic manipulation, we find that

$$\mathrm{erf}(z/\sqrt{2}) = t \left(1 - \mathrm{erf}(c/\sqrt{2}) \right) + \mathrm{erf}(c/\sqrt{2}).$$

Again, fortunately there are effective algorithms to calculate the inverse of the error function, for example erfinv in Matlab:

$$z = \Phi^{-1}(t) = \sqrt{2} \mathrm{erfinv} \left(t \left(1 - \mathrm{erf}(c/\sqrt{2}) \right) + \mathrm{erf}(c/\sqrt{2}) \right).$$

Hence, random generation in Matlab is very simple:

```
a = erf(c/sqrt(2));
t = rand;
z=sqrt(2)*erfinv(t*(1-a)+a);
```

Here, a warning is in order: if the bound c is large, the above program does not work. The reason is that the error function saturates quickly to unity, so when c is large, numerically $a = 1$, and the above algorithm gives numerical noise as a value of z. When working with large values of c, one needs to modify the code, but we do not discuss the modification here. Rather, we leave the modification as an exercise.

The Golden rule can be applied to integer valued densities, too, as the next example demonstrates.

EXAMPLE 5.3: Consider the problem of drawing from the Poisson distribution,

$$\pi(n) = \pi(n \mid \theta) = \frac{\theta^n}{n!} e^{-\theta}, \quad n = 0, 1, 2, \cdots,$$

whose cumulative distribution is also discrete,

$$\Phi(k) = e^{-\theta} \sum_{n=0}^{k} \frac{\theta^n}{n!}.$$

The random draw is done now via the following steps:
1. Draw $t \sim \text{Uniform}([0, 1])$,
2. Find the smallest integer k such that $\Phi(k) > t$.
The following is a segment of Matlab code that performs the random draw:

```
t = rand;
Phi = exp(-theta);
p = exp(-theta);
n = 0;
while Phi<=t
        p = p * (theta /(n + 1));
        Phi = Phi + p;
        n = n + 1;
end
k = n;
```

This algorithm was used in Chapter 3 to produce Figure 3.3.

5.3 Rejection sampling: prelude to Metropolis-Hastings

The Golden Rule can be easily extended to multivariate distributions provided that the components are mutually independent. If this is not the case, more advanced techniques need to be introduced. This leads us to consider the alternative (b) in the previous section, drawing from an easy *surrogate* distribution and then later correcting the sample in some way. As a prelude to this, we consider first a rather primitive idea.

EXAMPLE 5.4: Consider the distribution of Example 5.1, which is a Gaussian density with a bound constraint. The most straightforward approach for generating a sample from this distribution would probably be the following trial-and-error algorithm:

1. Draw x from the normal distribution $\mathcal{N}(0,1)$,
2. If $x > c$, accept, otherwise reject.

It turns out that this simple algorithm is a particular case of what is known as the *rejection sampling algorithm*. It works, but it can be awfully slow. To understand why, consider the *acceptance rate* for different values of c. If $c = 0$, in the average every other proposal will be accepted, since the underlying normal distribution is symmetric. If $c < 0$, the acceptance rate will be higher than 50%. On the other hand, if $c > 0$, it will be smaller. In general, the probability of hitting the domain of acceptance $\{x > c\}$ is

$$I(c) = \frac{1}{\sqrt{2\pi}} \int_c^\infty e^{-x^2/2} dx,$$

and $I(c) \to 0$ superexponentially as $c \to \infty$. For instance, if $c = 3$, the acceptance rate is less than 0.1%.

The situation gets worse for multidimensional problems. Consider for example a multivariate distribution in $\mathbb{R}^n$,

$$\pi(x) \propto \pi_+(x)\exp\left(-\frac{1}{2}|x|^2\right), \quad x \in \mathbb{R}^n,$$

where

$$\pi_+(x) = \begin{cases} 1, & \text{if } x_j > 0 \text{ for all } j, \\ 0, & \text{else} \end{cases}.$$

What is the acceptance rate using the trial-and-error algorithm? The normal distribution is symmetric around the origin, so any sign combination is equally probable. Since

$$P\{x_j > 0\} = \frac{1}{2}, \quad 1 \le j \le n,$$

the rate of successful proposals is

$$P\{x_1 > 0, x_2 > 0, \dots, x_n > 0\} = \prod_{j=1}^n P\{x_j > 0\} = \left(\frac{1}{2}\right)^n,$$

and it is clear that even the innocent looking positivity constraint puts us out of business if n is large. The central issue therefore becomes how to modify the initial random draws, step 1 above, to improve the acceptance.

Inspired by the trial-and-error drawing strategy of the previous example, we discuss next an elementary sampling strategy that works – at least in theory – also for multidimensional distributions.

Assume that the true probability density $\pi(x)$ is known up to a multiplicative constant, i.e., we have $C\pi(x)$ with $C > 0$ unknown, and we have a *proposal distribution*, $q(x)$ that is easy to use for random draws – for instance, a Gaussian distribution. Furthermore, assume that the proposal distribution satisfies the condition

$$C\pi(x) \le Mq(x) \quad \text{for some } M > 0.$$

The following algorithm is a prototype for the Metropolis-Hastings algorithm that will be introduced later:

1. Draw x from the proposal distribution $q(x)$,
2. Calculate the *acceptance ratio*

$$\alpha = \frac{C\pi(x)}{Mq(x)}, \quad 0 < \alpha \le 1.$$

3. Flip α–*coin*: draw $t \sim \text{Uniform}([0, 1])$, and accept x if $\alpha > t$, otherwise reject.

The last step says simply that x is accepted with probability α, rejected with probability $1 - \alpha$, thus the name α–coin.

Why does the algorithm work? What we need to show is that the distribution of the accepted x is π. To this end, let us define the event

$$\mathcal{A} = \text{event that a draw from } q \text{ is accepted},$$

regardless of what the draw is. The distribution of the accepted samples *drawn from q*, i.e., the distribution of the sample that we have just generated, is

$$\pi(x \mid \mathcal{A}) = \text{distribution of } x \text{ provided that it is accepted}.$$

To calculate this distribution, we use the Bayes formula,

$$\pi(x \mid \mathcal{A}) = \frac{\pi(\mathcal{A} \mid x)q(x)}{\pi(\mathcal{A})}. \tag{5.6}$$

Above, the prior of x is obviously q since we draw it from q. One interpretation of this is that before we test the acceptance, we believe that q is the correct distribution. What are the other densities appearing above?

The likelihood, the probability of acceptance *provided that x is given*, is clearly the acceptance rate,

$$\pi(\mathcal{A} \mid x) = \alpha = \frac{C\pi(x)}{Mq(x)}.$$

The marginal density of the acceptance is, in turn,

$$\pi(\mathcal{A}) = \int \pi(x, \mathcal{A})dx = \int \pi(\mathcal{A} \mid x)q(x)dx$$

$$= \int \frac{C\pi(x)}{Mq(x)}q(x)dx = \frac{C}{M}\int \pi(x)dx$$

$$= \frac{C}{M}.$$

Putting all the pieces in (5.6), we have

$$\pi(x \mid \mathcal{A}) = \frac{(C\pi(x)/Mq(x))q(x)}{C/M} = \pi(x),$$

exactly as we wanted.

We conclude this section by checking that the simple example of Gaussian distribution with a bound constraint is really a version of a rejection sampling algorithm. In that case, the true distribution, up to a normalizing constant, is

$$C\pi(x) = \pi_c(x)\exp\left(-\frac{1}{2}x^2\right),$$

the proposal distribution is Gaussian,

$$q(x) = \frac{1}{\sqrt{2\pi}}\exp\left(-\frac{1}{2}x^2\right),$$

and the scaling constant M can be chosen as $M = \sqrt{2\pi}$ so that

$$Mq(x) = \exp\left(-\frac{1}{2}x^2\right) \geq C\pi(x).$$

The acceptance rate is then

$$\alpha = \frac{C\pi(x)}{Mq(x)} = \pi_c(x) = \begin{cases} 1, & \text{if } x > c, \\ 0, & \text{if } x \leq c. \end{cases}$$

Flipping the α–coin in this case says that we accept automatically if $x > c$ ($\alpha = 1$) and reject otherwise ($\alpha = 0$), which is exactly what our heuristic algorithm was doing.

Exercises

1. Test numerically how large the parameter c in Example 5.2 may be without the algorithm breaking down. Implement an alternative algorithm along the following guidelines: select first an upper limit $M > c$ such that if $x > M$, $\pi(x)/\pi(c) < \delta$, where $\delta > 0$ is a small cut-off parameter. Then divide the interval $[c, M]$ in n subintervals, compute a discrete approximation for the cumulative distribution function and approximate numerically $\Phi^{-1}(t)$, $t \sim \text{Uniform}([0, 1])$. Compare the results with the algorithm of Example 5.2 when c is small.

2. Implement the rejection sampling algorithm when π is the truncated Gaussian density as in Example 5.2 and the proposal distribution an exponential distribution, $q(x) = \beta\exp(-\beta(x - c))$, $x \geq c$. Try different values for β and investigate the acceptance rate as a function of it.

6

Statistically inspired preconditioners

In Chapter 4 we discussed iterative linear system solvers and their application to ill-conditioned inverse problems, following a traditional approach: numerical linear algebra helps in solving problems in computational statistics. Characteristic of this approach is that when solving an ill-conditioned linear system $Ax = b$, the remedies for ill-posedness usually depend on the properties of the matrix A, not on what we expect the solution x to be like. It is interesting to change perspective, and devise a strategy for importing statistical ideas into numerical linear algebra. Statistically inspired preconditioners provide a method for doing this.

The choice of a particular iterative method for the solution of a linear system of equations is usually dictated by two considerations: how well the solution can be represented in the subspace where the approximate solution is sought, and how fast the method will converge. While the latter issue is usually given more weight in the general context of solving linear systems of equations, the former is more of interest to us. In fact, the subjective component in solving a linear system is *why* we are solving the particular system, and *what* we expect the solution to look like, which is the essence of Bayesian statistics.

The solution of a linear system by iterative solvers is sought in a subspace that depends on the matrix defining the system. If we want to incorporate prior information about the solution into the subspaces, evidently we have to modify the linear system using the prior information. This is the principal idea in this chapter.

We start by briefly reviewing some of the basic facts and results about preconditioners which we will be using later. Given a linear system of equations

$$Ax = b, \quad A \in \mathbb{R}^{n \times n}, \tag{6.1}$$

with A invertible, and a nonsingular matrix $M \in \mathbb{R}^{n \times n}$, it is immediate to verify that the linear system

$$M^{-1}Ax = M^{-1}b \tag{6.2}$$

has the same solution as (6.1). Likewise, if $R \in \mathbb{R}^{n \times n}$ is a nonsingular matrix, the linear system

$$AR^{-1}w = b, \quad Rx = w, \tag{6.3}$$

has also the same solution x as (6.1).

The *convergence rate* of an iterative method for the solution of a linear system of equations (6.1) depends typically on the spectral properties of the matrix A. Thus, if we replace the linear system (6.1) with (6.2) or (6.3), the rate of convergence will depend on the spectral properties of $M^{-1}A$ or AR^{-1}, respectively, instead of on the spectral properties of A. It is therefore clear that if M or R are well chosen, the convergence of an iterative method for (6.2) or (6.3), respectively, may be much faster than for (6.1).

The matrix M (R, respectively) is called a *left (right) preconditioner* and the linear system (6.2) (or (6.3)) is referred to as the *left (right) preconditioned linear system*. Naturally, it is possible to combine both right and left preconditioners and consider the system

$$M^{-1}AR^{-1}w = M^{-1}b, \quad Rx = w.$$

In the statistical discussion to ensue, the left and right preconditioners will have different roles: one will be related to the likelihood, the other to the prior.

6.1 Priorconditioners: specially chosen preconditioners

In general, the closer the matrix A is to the identity, the easier it becomes for a Krylov subspace iterative method to sufficiently reduce the norm of the residual error. Thus, for the construction of good preconditioners it is desirable to find a matrix M or R such that AR^{-1} or $M^{-1}A$ is close to an identity.

Since the application of an iterative method to a preconditioned linear system will require the computation of matrix-vector products of the form $M^{-1}Ax$ or $AR^{-1}w$, followed by the solution of linear systems of the form $Rx = w$ in the case of right preconditioners, it is also important that these computations can be done efficiently.

In the general case, the question of whether we should use left or right preconditioning depends only on the iterative method that we are using and on whether or not there are reasons to keep the right-hand side of the original linear system unchanged. Furthermore, the best general purpose preconditioners are those which yield the fastest rate of convergence with the least amount of work.

The situation becomes a little different when dealing with linear systems of equations with an ill-conditioned matrix and a right-hand side contaminated by noise. Since, as we have seen in Chapter 4, as the iteration number increases and amplified noise components start to corrupt the computed solution, it is important to avoid preconditioners which speed up the convergence of the noise.

In Chapter 4 we also saw that a linear system's sensitivity to noise was related to the smallest eigenvalues, or more generally, the smallest singular values. In an effort to accelerate the rate of convergence of the iterative methods, while keeping the noise in the right hand side from overtaking the computed solution, preconditioners which cluster only the largest eigenvalues of the matrix A, while leaving the smaller eigenvalues alone, have been proposed in the literature. The problem with this class of preconditioners is that the separation of the spectrum of A into the eigenvalues which should be clustered and those which should be left alone may be difficult if there is no obvious gap in the spectrum. Furthermore, the separation of the spectrum should depend on the noise level. Moreover, finding the spectral information of the matrix may be computationally challenging and costly.

One goal in the general preconditioning strategy is to gain spectral information about the operator and to modify it to obtain an equivalent system with better convergence properties. In our case, however, the main interest is in the properties of the unknowns, some of which we may know a priori, and the preconditioner is the Trojan horse which will carry this knowledge into the algorithm. Thus, instead of getting our inspiration about the selection of preconditioners from the general theory about iterative system solvers, we examine them in the light of linear Gaussian models that are closely related also to Tikhonov regularization.

The design of regularizing preconditioners starts traditionally from Tikhonov regularization, which is a bridge between statistical and non-statistical theory of inverse problems. Instead of seeking a solution to the ill-conditioned system (6.1), the strategy in Tikhonov regularization is to replace the original problem by a nearby minimization problem,

$$x_\delta = \arg\min \left(\|Ax - b\|^2 + \delta^2 \|Rx\|^2 \right), \tag{6.4}$$

where the penalty term is selected in such a way that for the desired solution, the norm of Rx is not excessively large, and the regularization parameter δ is chosen, e.g., by the discrepancy principle, if applicable.

The simplest version of Tikhonov regularization method – sometimes, erroneously, called *the* Tikhonov regularization method, while calling (6.4) a generalized Tikhonov regularization[1] – is to choose $R = I$, the identity matrix, leading to the minimization problem

$$x_\delta = \arg\min \left(\|Ax - b\|^2 + \delta^2 \|x\|^2 \right). \tag{6.5}$$

As we have indicated in Chapter 4, an alternative to Tikhonov regularization (6.5) is to use iterative solvers with early truncation of the iterations. In particular, when the CGLS method is employed, the norms of the iterates

---

[1] Andrei Nikolaevich Tikhonov (1906–1993), who had a great impact on various areas of mathematics, was originally a topologist, and, loyal to his background, thought of regularization in terms of compact embeddings.

form a non-decreasing sequence, yielding a straightforward way of monitoring the penalty term $\|x\|$.

So what could be an alternative to the more general regularization strategy (6.4)? It is obvious that if R is a nonsingular square matrix, the regularization strategy (6.4) is equivalent to

$$w_\delta = \arg\min\left(\|AR^{-1}w - b\|^2 + \delta^2\|w\|^2\right), \quad Rx_\delta = w_\delta,$$

and the natural candidate for an alternative strategy to Tikhonov regularization would be a truncated iterative solver with right preconditioner R.

To understand the effects of choosing preconditioners in this fashion, we go back to the Bayesian theory, which was our starting point to arrive at Tikhonov regularization. We consider the linear additive model

$$B = AX + E, \quad E \sim \mathcal{N}(0, \sigma^2 I), \quad X \sim \mathcal{N}(0, \Gamma),$$

whose posterior probability density is

$$\pi(x \mid b) \propto \exp\left(-\frac{1}{2\sigma^2}\|b - Ax\|^2 - \frac{1}{2}x^\mathrm{T}\Gamma^{-1}x\right).$$

Let

$$\Gamma = UDU^\mathrm{T},$$

be the eigenvalue decomposition of the covariance matrix Γ, where U is an orthogonal matrix whose columns are the eigenvectors of Γ and the diagonal matrix

$$D = \mathrm{diag}(d_1, d_2, \ldots, d_n), \quad d_1 \geq d_2 \geq \cdots \geq d_n,$$

contains the eigenvalues in decreasing order. After introducing the symmetric factorization

$$\Gamma^{-1} = UD^{-1/2}\underbrace{D^{-1/2}U^\mathrm{T}}_{=R} = R^\mathrm{T}R,$$

we can write the posterior in the form

$$\pi(x \mid b) \propto \exp\left(-\frac{1}{2\sigma^2}\left[\|b - Ax\|^2 + \sigma^2\|Rx\|^2\right]\right),$$

from which it can be easily seen that the Maximum A Posteriori estimate is the Tikhonov regularized solution with $\delta = \sigma$ and

$$R = D^{-1/2}U^\mathrm{T}. \tag{6.6}$$

This formulation lends itself naturally to a geometric interpretation of Tikhonov penalty. Since

$$Rx = \sum_{j=1}^{n}\frac{1}{\sqrt{d_j}}(u_j^\mathrm{T}x),$$

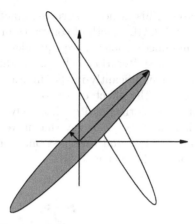

Fig. 6.1. The light ellipse is the preimage of a disc in the mapping A, the darker one is an equiprobability ellipse defined by the prior. The eigenvectors of the prior covariance matrix are also shown.

the penalty term effectively pushes the solution towards the eigenspace corresponding to *largest* eigenvalues of the covariance matrix by penalizing the solution for the growth of those components $u_j^T x$ that are divided by small numbers $\sqrt{d_j}$.

EXAMPLE 6.1: We illustrate graphically what is going on, considering the 2×2 toy example discussed in Examples 4.5 and 4.6, equipped with a Gaussian prior. In Figure 6.1 we have plotted the preimage of a small disc around the noiseless data (light ellipse), as in Example 4.5, and in addition a shaded equiprobability ellipse of the prior density. By the very definition of prior, we believe that the solution could a priori have large variance in the direction of the major semiaxis of this ellipse, which is also the eigendirection corresponding to the larger eigenvalue d_1 of the covariance matrix, while the variance should be small in the direction of the minor semiaxis, which is the eigendirection corresponding to the smaller eigenvalue d_2.

We solve the original linear system

$$Ax = b_j, \quad b_j = b_* + e_j, \quad 1 \le j \le N,$$

with a sample of noisy right hand sides, with the noise realizations e_j drawn from a Gaussian white noise density. We choose to solve the problem iteratively, using CGLS iteration and ignoring the prior. We then repeat the process, this time solving the preconditioned linear system

$$AR^{-1}w = b_j, \quad Rx = w, \quad b_j = b_* + e_j, \quad 1 \le j \le N,$$

with the same data set and using CGLS iteration. The results are shown in Figures 6.2–6.3. We observe that after two iterations the solutions are the same in both cases. This is not a surprise, since for a regular two-dimensional problem, the CGLS method converges in two iterations, and the original and the modified problems are equivalent. Differences are visible, however, after the first iteration. The preconditioning reduced the variability of the solution significantly in the direction of the small eigenvalue of the prior. Hence, the effect of preconditioning was to produce approximate solutions that correspond qualitatively to what we believed *a priori* of the solution. Thus, we may say that if we wanted to solve the problem with truncated CGLS and take just one iteration step, qualitatively the use of the preconditioner paid off.

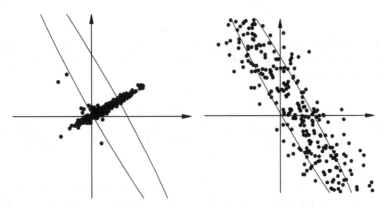

Fig. 6.2. Approximate solutions of the linear systems in Example 6.1, corresponding to a sample of 200 noisy realizations of the right hand side, obtained after one (left) and two (right) iterations of the CGLS method.

Inspired by the small example discussed above, consider now a general linear model with Gaussian prior and Gaussian additive noise,

$$B = AX + E, \quad X \sim \mathcal{N}(x_0, \Gamma), \quad E \sim \mathcal{N}(e_0, \Sigma),$$

and X and E mutually independent. After introducing a symmetric factorization, e.g. the Cholesky decompositions, of the inverses of the covariance matrices,

$$\Gamma^{-1} = R^T R, \quad \Sigma^{-1} = S^T S, \tag{6.7}$$

the posterior density can be expressed as

$$\pi(x \mid b)$$

$$\propto \exp\left(-\frac{1}{2}(b - Ax - e_0)^T \Sigma^{-1}(b - Ax - e_0) - \frac{1}{2}(x - x_0)^T \Gamma^T (x - x_0) \right)$$

$$= \exp\left(-\frac{1}{2}\left[\|S(b - Ax - e_0)\|^2 + \|R(x - x_0)\|^2 \right] \right),$$

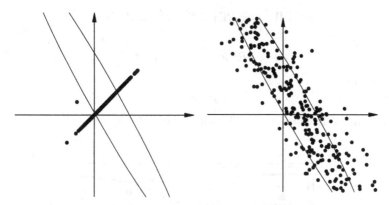

Fig. 6.3. Approximate solutions of the linear systems in Example 6.1, corresponding to a sample of 200 noisy realizations of the right hand side, obtained after one (left) and two (right) iterations of the CGLS method, using the whitening matrix of the prior as priorconditioner.

and by defining

$$w = R(x - x_0), \tag{6.8}$$

we observe that

$$\|S(b - Ax - e_0)\|^2 + \|R(x - x_0)\|^2 = \|y - SAR^{-1}w\|^2 + \|w\|^2,$$

where

$$y = S(b - Ax_0 - e_0). \tag{6.9}$$

The derivation above suggests the following regularized algorithm for estimating the solution x:

1. Calculate the factorizations (6.7) of the noise and prior covariances, and the vector (6.9).
2. Approximate the solution of the transformed system

$$SAR^{-1}w = y, \quad R(x - x_0) = w$$

using a suitable iterative solver.

We call the matrices S and R the *left and right priorconditioners*. Observe that the left priorconditioner is related to the structure of the noise, while the right priorconditioner carries information from the prior.

We emphasize that the priorconditioners are the whitening operators of X and E, respectively. If, for simplicity, we assume that $x_0 = 0$, $e_0 = 0$, and let

$$W = RX, \quad Y = SE,$$

then

$$E\{WW^{\mathrm{T}}\} = RE\{XX^{\mathrm{T}}\}R^{\mathrm{T}} = R\Gamma R^{\mathrm{T}}.$$

Since

$$\Gamma = (R^{\mathrm{T}}R)^{-1} = R^{-1}R^{-\mathrm{T}},$$

it follows that

$$R\Gamma R^{\mathrm{T}} = I,$$

showing that W is white noise. The same is true for Y.

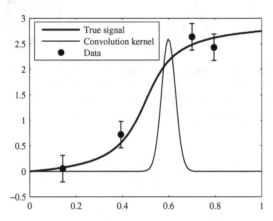

Fig. 6.4. The true signal, convolution kernel and the four data points. The error bars indicate plus/minus two standard deviations of the noise in the data.

EXAMPLE 6.2: In this example, we consider the deconvolution of a signal from very sparse and noisy data. Suppose that we have a continuous signal f supported on the unit interval $[0, 1]$ with $f(0) = 0$, of which we know four blurred and noisy samples,

$$g(s_j) = \int_0^1 a(s_j - t)f(t)dt + e_j, \quad 1 \le j \le 4.$$

The relatively narrow Gaussian convolution kernel is shown with the data and signal that was used to generate it in Figure 6.4. The task is to estimate the signal f from this data.

After discretizing the problem by subdividing the interval $[0, 1]$ into n equal subintervals and letting $x_k = f(t_k)$, where t_k is one of the division points, we obtain the linear system

$$b = Ax + e,$$

where $b \in \mathbb{R}^4$, $b_j = g(s_j)$, represents the data. Here we choose $n = 128$, leading to a badly underdetermined system.

Since the matrix $A \in \mathbb{R}^{4 \times 128}$ is non-square, of the three iterative methods discussed in Section 4, only the CGLS method can be applied. For comparison, two versions, one with and one without a priorconditioner are tested. Therefore, let us first define a prior. In this example, the prior belief is that the true signal is not oscillating excessively, and that $f(0) = 0$. A stochastic model expressing our belief that the value at $t = t_k$ cannot be very different from the value at $t = t_{k-1}$, can be expressed as

$$X_k = X_{k-1} + W_k, \quad 1 \le k \le n, \quad X_0 = 0. \tag{6.10}$$

We model the uncertainties W_k by mutually independent zero mean Gaussian random variables, whose standard deviations express how much we believe that the adjacent values may differ from one another. If we believe that the range of the signal values are of the order one, a reasonable value for the standard deviations is of the order $1/n$. The equation (6.10) can be written in the matrix form as

$$RX = W,$$

where the matrix R is the first order finite difference matrix

$$R = \begin{bmatrix} 1 & & & \\ -1 & 1 & & \\ & \ddots & \ddots & \\ & & -1 & 1 \end{bmatrix} \in \mathbb{R}^{n \times n}.$$

Assuming that the standard deviations of the variables W_k are all equal to γ, we arrive at a first order smoothness prior model,

$$\pi_{\text{prior}}(x) \propto \exp\left(-\frac{1}{2\gamma^2} \|Rx\|^2 \right),$$

Since the matrix R is invertible, it can be used as priorconditioner.

In Figure 6.5, we have plotted the first four iterations of the CGLS algorithm with and without priorconditioner. The effect of the priorconditioning is evident. The eigenvectors of the prior covariance matrix associated with large eigenvalues correspond to signals with small jumps, i.e., Rx is small. Therefore the priorconditioned solutions show fewer oscillations than those obtained without priorconditioning.

In the light of the example above, it is worthwhile to make a small comment concerning the usefulness of priorconditioning. In the context of Bayesian models for linear Gaussian random variables, the Maximum A Posteriori estimate x_{MAP} satisfies

$$x_{\text{MAP}} = \arg\min \left\| \begin{bmatrix} A \\ \delta R \end{bmatrix} x - \begin{bmatrix} b \\ 0 \end{bmatrix} \right\|, \quad \delta = \frac{\sigma}{\gamma},$$

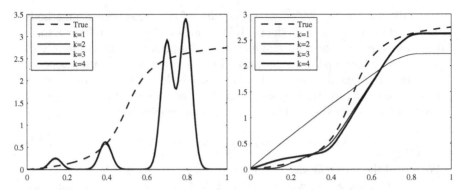

Fig. 6.5. The first few CGLS iterations without the priorconditioner (left) and with the preconditioner (right). Observe that, without the priorconditioner, the approximate solution after one iteration does not change anymore, thus the solutions corresponding to different iterations are indistinguishable. The priorconditioned approximate solutions seem to stabilize after three iterations.

which might lead us to argue that the correct Bayesian estimator is the least squares solution of the linear system

$$\begin{bmatrix} A \\ \delta R \end{bmatrix} x = \begin{bmatrix} b \\ 0 \end{bmatrix}, \tag{6.11}$$

which can be computed using an iterative solver. This argument is correct, of course, with the *caveat* that while the priorconditioned CGLS does not require that the value of γ, or, equivalently, of δ, is known, the MAP estimate does. If we do not know it, its estimation becomes part of the Bayesian inverse problem and requires additional work. This is also the case if we interpret δ as Tikhonov regularization parameter. Therefore we conclude that the potentially difficult step of selecting the value of the prior parameter is finessed by priorconditioning, where it is traded with the selection of a truncation index.

Another worthwhile comment concerns the dimensions of the matrices and the available selection of iterative solvers. Assuming that the matrix A is square, the priorconditioned problem can be solved by the GMRES method, which is, in many cases, more efficient than the CGLS method. If, however, instead of the priorconditioned problem we decide to solve the MAP estimate from (6.11), the selection of applicable iterative solver is narrower.

So far, we have only discussed right preconditioning. It is interesting to consider also the advantages of left preconditioning. For this purpose we return to our two-dimensional problem.

EXAMPLE 6.3: Consider a 2×2 linear Gaussian model,

$$B = Ax_* + E, \quad E \sim \mathcal{N}(0, \Sigma).$$

where the matrix A is as in Examples 4.5 and 4.6, but instead of being Gaussian white, the additive noise has a covariance matrix different from

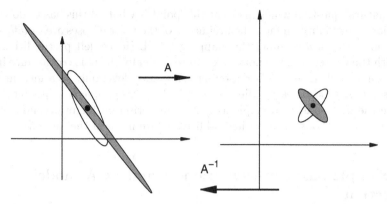

Fig. 6.6. Two different equiprobability ellipses of the data (right) and the corresponding preimages in the mapping A (left).

unity. With additive Gaussian white noise, the equiprobability curves of B are circles around the mean Ax_* because the variance of the noise is the same in all directions. In this example, we assume that the covariance matrix of the noise has a spectral factorization of the form

$$\Sigma = V\mathrm{diag}(s_1^2, s_2^2)V^{\mathrm{T}},$$

where V is a matrix whose columns are the eigenvectors of the matrix Σ scaled to have unit length. Due to the symmetry of the covariance matrix, the eigenvectors are orthogonal and we can write V in the form

$$V = \begin{bmatrix} \cos\varphi & -\sin\varphi \\ \sin\varphi & \cos\varphi \end{bmatrix}.$$

If we assume that $s_1 > s_2$, the equiprobability curves of B are no longer circles but ellipses, with principal semi-axes of length s_1 and s_2 whose orientations depend on the angle φ.

In the right panel of Figure 6.6, we have plotted two equiprobability ellipses, corresponding to $s_1 = 1$, $s_2 = 0.2$, and $\varphi = \pi/4$ in the first (lighter ellipse) and $\varphi = 3\pi/4$ in the second case (darker ellipse). The corresponding preimages of these ellipses under the mapping A are shown in the left panel of same figure. We conclude that by changing the angle φ, which does not change the noise level but only its covariance structure, the sensitivity of the inverse problem to observation noise changes significantly. In fact, the (darker) preimage of the noise ellipse corresponding to $\varphi = 3\pi/4$ is remarkably more elongated than the one corresponding to $\varphi = \pi/4$, indicating that in one direction the sensitivity to noise has been increased significantly. This example demonstrates that it is certainly not enough to look at the noise level alone, as is customary in the non-statistical theory, but that the covariance, and more generally the probability density of the noise is also very important.

A natural question which arises at this point is what all this has to do with priorconditioning. In fact, the preimages of the noise ellipses are *exactly* the preimages of a *disc* under the mapping $S^{-1}A$. Hence, left preconditioning with the whitener of the noise takes automatically the noise covariance into account, and the classical reasoning using noise level only as a measure of discrepancy becomes justified. In particular, this simple step justifies the use the discrepancy principle as a stopping criterion for priorconditioned iterative methods, as if we had additive white noise in the model.

6.2 Sample-based preconditioners and PCA model reduction

The discussion above suggests how to construct preconditioners when information about the statistics of the solution and of the noise is available. In many applications, however, instead of the statistics of the solution we have access to a collection of typical solutions. A central question in statistical modelling of inverse problems is how to construct informative and reasonable prior densities. Here we discuss the problems of the estimation of prior densities from available samples and we warn about overly committal priors that are biasing towards a reasonable but too conservative solution. In fact, while the prior should favor the typical or normal solutions that we are expecting to see, at the same time it should not exclude the presence of abnormalities or anomalies that might in fact be, e.g., in medical imaging, what we are really interested in.

The problem that we are considering here is how to *estimate a quantity x from observations b that are related via a computational model.* Assume that we have access to a sample believed to consist of realizations of the random variable X representing the unknown of primary interest, as well as to the corresponding set of data, believed to be realizations of the observable random variable B. We call the sample of corresponding pairs a *training set* and denote it by

$$S_0 = \big\{(x_1, b_1), (x_2, b_2), \ldots, (x_N, b_N)\big\}, \quad x_j \in \mathbb{R}^n, \; b_j \in \mathbb{R}^m,$$

with $N \geq n$. The noise level of the data may or may not be known. The set S_0 may have been constructed from simulations or it may be the result of actual experimental recordings. In medical applications, for example, the learning set could consist of indirect, noninvasive measurements b_j combined with information obtained during subsequent surgical interventions. We assume that the vectors x_j are discretized approximations of the quantities of interest, and that the discretization level is fixed and determined by the grid in which we seek to solve the inverse problem.

If the sample is representative and N is sufficiently large, it is possible to estimate the probability density of the underlying variable X using this sample. The problem of estimating a probability density based on a sample is the classical problem of frequentist statistics, so our approach has a non-Bayesian flavor.

Given a sample of true realizations, it may be unrealistic to assume that they are normally distributed, and in fact, the validity of such assumption may be tested. However, to construct priorconditioners we seek a Gaussian *approximation* of the prior density. The Gaussian distributions are completely characterized by the mean and the covariance. Estimates of the mean and the covariance of X can be obtained from the available sample, as

$$x_* = \mathrm{E}\{X\} \approx \frac{1}{N}\sum_{j=1}^{N} x_j = \tilde{x}_*,$$

$$\Gamma = \mathrm{E}\{XX^\mathrm{T}\} - x_* x_*^\mathrm{T} \approx \frac{1}{N}\sum_{j=1}^{N} x_j x_j^\mathrm{T} - \tilde{x}_* \tilde{x}_*^\mathrm{T} = \tilde{\Gamma}. \qquad (6.12)$$

Higher order moments can be estimated as well, and they can be used to assess the fidelity of the Gaussian approximation.

In the applications that we have in mind, the vectors x_j represent typical features of the random variable X. This often means that the vectors are not very dissimilar. Consequently, the space spanned by the realizations may be a proper subspace even if $N \geq n$, thus $\tilde{\Gamma}$ may be rank deficient or of ill-determined rank[2].

Assume first that $\tilde{\Gamma}$ is a satisfactory approximation of Γ. Without loss of generality, we may assume that the mean of X vanishes. Let

$$\tilde{\Gamma} = VDV^\mathrm{T}, \quad V = [v_1, v_2, \ldots, v_n], \quad D = \mathrm{diag}\,(d_1, d_2, \ldots, d_n), \qquad (6.13)$$

be the singular value decomposition of the matrix $\tilde{\Gamma}$. Assume that only the first r singular values are larger that a preset tolerance, and set the remaining ones to zero, i.e., $d_1 \geq d_2 \geq \ldots \geq d_r > d_{r+1} = \ldots = d_n = 0$ and partition V accordingly,

$$V = \begin{bmatrix} V_0 & V_1 \end{bmatrix}, \quad V_0 = [v_1, \ldots, v_r], \quad V_1 = [v_{r+1}, \ldots, v_n].$$

By using the orthogonality of V, we have

$$I = VV^\mathrm{T} = \begin{bmatrix} V_0 & V_1 \end{bmatrix} \begin{bmatrix} V_0^\mathrm{T} \\ V_1^\mathrm{T} \end{bmatrix} = V_0 V_0^\mathrm{T} + V_1 V_1^\mathrm{T},$$

and therefore we may split X as

---

[2] By ill-determined rank we mean that the eigenvalues, or more generally, singular values, of the matrix go to zero, but there is no obvious cutoff level below which the eigenvalues could be considered negligible. This type of ill-posedness is often very difficult to handle in numerical analyis.

$$X = V_0(V_0^{\mathrm{T}}X) + V_1(V_1^{\mathrm{T}}X) = V_0 X_0 + V_1 X_1, \quad X_0 \in \mathbb{R}^r, \; X_1 \in \mathbb{R}^{n-r}.$$

To estimate the expected size of X_1, we write

$$\mathrm{E}\{\|X_1\|^2\} = \sum_{j=1}^{n} \mathrm{E}\{(X_1)_j^2\} = \mathrm{trace}\left(\mathrm{E}\{X_1 X_1^{\mathrm{T}}\}\right),$$

and further, by expressing X_1 in terms of X we have

$$\mathrm{E}\{X_1 X_1^{\mathrm{T}}\} = V_1^{\mathrm{T}}\mathrm{E}\{XX^{\mathrm{T}}\}V_1 = V_1^{\mathrm{T}} \Gamma V_1.$$

By approximating $\Gamma \approx \widetilde{\Gamma}$ and substituting the right hand side of (6.13), it follows that

$$\begin{aligned}
\Gamma V_1 &= \Gamma \begin{bmatrix} v_{r+1} & v_{r+2} & \cdots & v_n \end{bmatrix} \\
&= \begin{bmatrix} \Gamma v_{r+1} & \Gamma v_{r+2} & \cdots & \Gamma v_n \end{bmatrix} \\
&= \begin{bmatrix} d_{r+1} v_{r+1} & d_{r+2} v_{r+2} & \cdots & d_n v_n \end{bmatrix},
\end{aligned}$$

and, from the orthogonality of the vectors v_j,

$$\mathrm{E}\{\|X_1\|^2\} = \mathrm{trace} \begin{bmatrix} d_{r+1} & & \\ & \ddots & \\ & & d_n \end{bmatrix} = \sum_{j=r+1}^{n} d_j = 0,$$

which is equivalent to saying that, within the validity of the approximation of Γ by $\widetilde{\Gamma}$, $X = V_0 X_0$ with probability one. Therefore, we may write a *reduced model* of the form

$$B = AX + E = \underbrace{AV_0}_{=A_0} X_0 + E \qquad (6.14)$$

$$= A_0 X_0 + E, \quad A_0 \in \mathbb{R}^{m \times r}, \quad X_0 \in \mathbb{R}^r,$$

whose model reduction error has zero probability of occurrence. The above model is equivalent to the *Principal Component Analysis* (PCA) model[3]. Principal Component Analysis is a popular and often effective way of reducing the dimensionality of a problem when it is believed that lots of the model parameters are redundant.

To better understand the properties of the reduced model, we introduce the matrix

$$D_0 = \mathrm{diag}\left(d_1, d_2, \ldots, d_r\right) \in \mathbb{R}^{r \times r},$$

and define the new random variable

$$W_0 = D_0^{-1/2} X_0 = D_0^{-1/2} V_0^{\mathrm{T}} X \in \mathbb{R}^r,$$

---

[3] See, e.g., [Jo02].

where
$$D_0^{-1/2} = \left(D_0^{1/2}\right)^{-1} = \mathrm{diag}\left(d_1^{-1/2}, d_2^{-1/2}, \ldots, d_r^{-1/2}\right).$$

The covariance matrix of the variable W_0 is then
$$\mathrm{E}\{W_0 W_0^{\mathrm{T}}\} = D_0^{-1/2} V_0^{\mathrm{T}} \Gamma V_0 D_0^{-1/2}.$$

Since
$$V_0^{\mathrm{T}} \Gamma V_0 = D_0,$$

it follows that
$$\mathrm{cov}\left(W_0\right) = I_{r \times r},$$

i.e., W_0 is r–variate white noise. The *whitened PCA model* for X can then be written as
$$B = A V_0 D_0^{1/2} W_0 + E, \quad X = V_0 D_0^{1/2} W_0, \tag{6.15}$$

where the matrix $V_0 D_0^{1/2} \in \mathbb{R}^{n \times r}$ acts as a whitening preconditioner. Notice that here we use the term preconditioner rather loosely, since the matrix is not even square unless $r = n$.

This formulation of the PCA model emphasizes the fact that

$$X = V_0 D_0^{1/2} W_0,$$

being a linear combination of the columns of the matrix V_0, is automatically in the subspace spanned by the singular vectors $v_1, \ldots, v_r$. Hence, the PCA model is a *hard subspace constraint*, since it forces the solution into a low dimensional space.

The obvious appeal of the PCA reduced model (6.14) in either its original or whitenend version (6.15), is that it takes full advantage of the prior information. When $r \ll n$, the computational work may decrease substantially. It should be pointed out, however, that the subspace spanned by V_0 may contain singular vectors of A that correspond to small singular values, thus the PCA reduced model may still exhibit high sensitivity to noise.

From the point of view of applications, the PCA model reduction is not void of problems. The first question which arises is related to the approximation $\Gamma \approx \tilde{\Gamma}$. In general, it is very difficult to assess to what extent the training set is a sufficient sample. If the vectors x_j are drawn independently from the same density, by the Central Limit Theorem the approximation error of Γ is asymptotically Gaussian with variance decreasing as $\mathcal{O}(1/N)$. However, the error associated with a given N can perturb significantly the singular values used to determine the truncation parameter r and, consequently, the PCA subspace may be unable to represent essential features of interest.

A more severe problem, certainly from the point of view of medical applications, is the PCA reduced model inability to reproduce *outliers*. Assume, for instance, that the training set represents thoracic intersection images. The majority of the images corresponds to normal thoraxes, while the few outliers representing anomalies, tumors, for example, that might be in the set have a negligible effect in the averaging process. As a consequence, the anomalous features will not be represented by the PCA subspace vectors, when the purpose of the imaging process might have been, in fact, *to detect these anomalies*. So we might have a perfect method of reproducing something that we know, but the essential information that we are looking for is discarded! One possible way to overcome this problem is discussed in the following example.

EXAMPLE 6.4: Assume that we believe a priori that the unknown consists of a regular part and an anomalous part that may or may not be present. We write a stochastic model,

$$X = X_r + V X_a, \tag{6.16}$$

where X_r and X_a are the anomalous and regular parts, respectively, and V is a random variable that takes on values zero or one. We assume that X_r and X_a are mutually independent. Furthermore, assume that

$$E\{X_r\} = \overline{x}_r, \quad E\{X_a\} = \overline{x}_a,$$

and

$$\mathrm{Cov}(X_r) = \Gamma_r, \quad \mathrm{Cov}(X_a) = \Gamma_a.$$

The probability of the occurrence of the anomaly is assumed to be α, so

$$P\{V = 1\} = \alpha, \quad P\{V = 0\} = 1 - \alpha.$$

The mean and the covariance matrices may have been estimated based on an empirical sample. The role of the anomalous part in the model is to avoid a too committal prior that leaves out interesting details from the estimate.

To apply the priorconditioning technique, we calculate a Gaussian approximation for the prior density of X. Therefore, the mean and the covariance need to be calculated. We use the discrete version of the formula (1.4) and write

$$E\{X\} = E\{X \mid V = 1\}P\{V = 1\} + E\{X \mid V = 0\}P\{V = 0\}$$

$$= (\overline{x}_r + \overline{x}_a)\alpha + \overline{x}_r(1 - \alpha) = x_r + \alpha x_a = \overline{x}.$$

Similarly, we calculate the covariance. In the calculation below, we assume for simplicity that $\overline{x}_a = 0$, leading to

$$\mathrm{Cov}(X) = E\{(X - x_r)(X - x_r)^T\}$$

$$= E\{(X - x_r)(X - x_r)^T \mid V = 1\}P\{V = 1\}$$

$$+ E\{(X - x_r)(X - x_r)^T \mid V = 0\}P\{V = 0\}$$

$$= \alpha E\{(X_r + X_a - x_r)(X_r + X_a - x_r)^T\}$$
$$+ (1 - \alpha)E\{(X_r - x_r)(X_r - x_r)^T\}$$
$$= \alpha(\Gamma_r + \Gamma_a) + (1 - \alpha)\Gamma_r$$
$$= \Gamma_r + \alpha\Gamma_a.$$

It is left as an exercise to show that if the mean of the anomalous part is non-vanishing, the covariance is

$$\text{Cov}(X) = \Gamma_r + \alpha\Gamma_a + \alpha(1 - \alpha)\bar{x}_a\bar{x}_a^T. \tag{6.17}$$

Consider now a deconvolution example where we estimate $f : [0,1] \to \mathbb{R}$ from

$$g(s_j) = \int_0^1 a(s_k - t)f(t)dt + e_j, \quad 0 \le s_j \le 1,$$

where $s_j = (j - 1)/(n - 1)$, $1 \le j \le n$ and $n = 200$. The kernel a is Gaussian and the noise vector is zero mean, normally distributed white noise. We discretize the problem using grid points that coincide with the data points s_j, leading to the linear model

$$b = Ax + e.$$

We assume a priori that the true signal f, defined over the unit interval $[0,1]$ is under regular conditions a boxcar function, and we have an interval estimate for both its height and location,

$$f_r(t) = h\chi_{[a,b]}(t),$$

where $\chi_{[a,b]}$ is a characteristic function of the interval $[a, b]$, and

$$1 \le h \le 1.2, \quad 0.2 \le a \le 0.25, \quad 0.65 \le b \le 0.7.$$

To produce the regular part of the mean and covariance matrix, we generate a sample of 20 000 vectors x that are discretizations of boxcar functions f_r above,

$$h \sim \text{Uniform}([1, 1.2]), \quad a \sim \text{Uniform}([0.2, 0.25]), \quad b \sim \text{Uniform}([0.65, 0.7]).$$

We then assume that the regular part may be contaminated by a rare anomaly of small support, and this anomaly is likely to appear in the non-vanishing part of f_r. Therefore, we write a model for the anomaly,

$$f_a(t) = C\exp\left(-\frac{1}{2\gamma^2}(t - t_0)^2\right),$$

and to generate a sample of anomalies, we assume probability distributions

$$C \sim \text{Uniform}([0.1, 0.5]), \quad t_0 \sim \text{Uniform}([0.4, 0.5]), \quad \gamma \sim \text{Uniform}([0.005, 0.03]).$$

Based on a sample of 20 000 vectors of discretizations of Gaussians with parameters drawn from these distribution, we calculate the mean and covariance of the anomalous part.

We now compute the eigenvalue decomposition of the covariance of the regular part,

$$\Gamma_{\mathrm{r}} = U_{\mathrm{r}} D_{\mathrm{r}} U_{\mathrm{r}}^{\mathrm{T}}.$$

The matrix Γ_{r} is of ill-determined rank. We discard the eigenvalues that are below a threshold $\tau = 10^{-10}$ and replace them by zeros. Let $\tilde{D}_{\mathrm{r}}$ denote the thresholded diagonal matrix. We then have

$$\Gamma_{\mathrm{r}} \approx U_{\mathrm{r}} \tilde{D}_{\mathrm{r}} U_{\mathrm{r}}^{\mathrm{T}} = \underbrace{(U_{\mathrm{r}} \tilde{D}_{\mathrm{r}}^{1/2})}_{= R_{\mathrm{r}}} (\tilde{D}_{\mathrm{r}}^{1/2} U_{\mathrm{r}}^{\mathrm{T}}) = R_{\mathrm{r}} R_{\mathrm{r}}^{\mathrm{T}}.$$

Similarly, we define the covariance matrix Γ defined by formula (6.17), with $\alpha = 0.5$, and write the decomposition

$$\Gamma = RR^{\mathrm{T}}.$$

Observe that if W is a white noise vector, then the random variable defined by

$$X = \overline{x} + RW$$

is normally distributed with mean $\overline{x}$ and covariance Γ. This leads us to consider the priorconditioned problem of solving the equation

$$b - A\overline{x} = ARw, \quad x = \overline{x} + Rw,$$

and we may apply iterative solvers with the discrepancy principle as a stopping criterion.

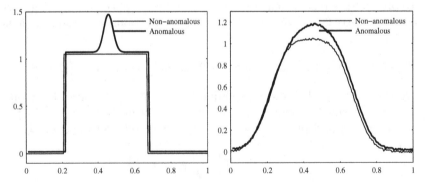

Fig. 6.7. The true signals used for data generation (left) and the noisy data. The signals are offset slightly to improve the visualization.

To test the idea, we generate two signals, one with and one without the anomaly, and calculate the corresponding noisy data. The signals and the data are shown in Figure 6.7. The additive noise has a standard deviation 0.5% of the maximum value of the noiseless signal that would correspond to the non-anomalous mean $\overline{x}_r$.

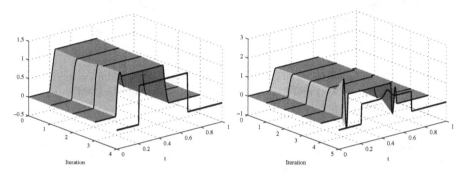

Fig. 6.8. The evolution of the CGLS iterations with the priorconditioner based on the learning sample with no anomalies and with the data coming from a non-anomalous source (left) and from a source with anomaly (right). The true signals are plotted on the foreground.

We test the priorconditioned CGLS algorithm in four different cases. First, we use only the non-anomalous part of the covariance matrix and test the algorithm both with the non-anomalous and the anomalous data. The results are shown in Figure 6.8. Evidently, the performance of the algorithm is good when the data comes from a non-anomalous source, while there is a strong edge effect visible when the data comes from the anomalous signal.

We than add the anomalous part to the covariance and run again the algorithm both with the anomalous and non-anomalous data. The results can be seen in Figure 6.9. Evidently, the addition of the anomalous part improves the performance with the anomalous data, but as importantly, it does not produce significant "false positive" artifact in the solution corresponding to the non-anomalous data.

In the previous example, the parameter α represents the probability of occurrence of an anomaly which may be a rare event, e.g., the occurrence of a tumor in a radiograph, so one might think that it should be chosen very small, unlike in the example, and the effect of the anomalous part must therefore become negligible. However, we must bear in mind that the prior is based on subjective judgment. If a patient is sent for an X–ray, it is likely that there is already a suspicion of something abnormal in the radiograph. Hence, the prior density is a conditional density, and the expectation of the anomaly may well be taken rather high.

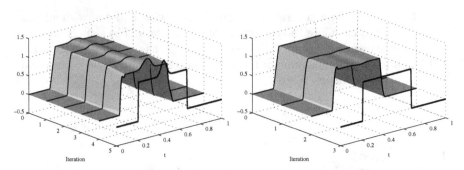

Fig. 6.9. The priorconditioner contains an anomalous part, the data coming from an anomalous source (left) and from a non-anomalous one (right). The true signals are plotted on the foreground.

Exercises

1. Given a n–variate random variable X with mean $\overline{x} \in \mathbb{R}^n$, show that

$$\mathrm{E}\{(X - a)(X - a)^{\mathrm{T}}\} = \mathrm{Cov}(X) + (\overline{x} - a)(\overline{x} - a)^{\mathrm{T}},$$

   where $a \in \mathbb{R}^n$ is any vector. This is the off-centered covariance of the variable X.
2. Consider the Example 6.4 when the anomalous part has a non-vanishing mean $\overline{x}_{\mathrm{a}}$. Using the result of Exercise 1, show that in this case, the variance of X is given by the formula (6.17).

7

Conditional Gaussian densities and predictive envelopes

From the computational point of view, a fundamental problem in Bayesian statistics is how to move between the marginal, conditional and joint probability distributions. To demonstrate various ways of doing this, we discuss some classical problems from the statistical point of view, and show that not only is the Bayesian analysis able to give the classical solutions, but it also provides means of assessing how confident we should feel about them, in the light of the data and our prior belief. In many applications, the latter aspect is important enough to provide a reason to switch to the Bayesian framework.

In the context of linear inverse problems with additive Gaussian noise and Gaussian prior, we observed that the joint probability density and the posterior density are also Gaussian. In the derivation of the posterior density we assumed that the noise and the unknown to be estimated were mutually independent. This assumption is fairly common, and it can often be justified, for example, when the source of the noise is exogenous. There are important cases, however, when this assumption does not hold, and the concept of noise should be defined in more general terms. The following examples are meant to clarify what we mean by this.

> EXAMPLE 7.1: Assume that we are recording electric or magnetic fields outside a person's skull, trying to infer on the cortical activity. The subject is shown images and the resulting electromagnetic activity on the visual cortex is the unknown of primary interest. Inevitably, when the images are shown, the subject moves the eyes, thereby causing activity on the motor cortex. The muscles that control the eye movements also produce a weak electromagnetic signal. Since these events are evoked by the showing of the images, we cannot assume that the highly correlated signals from the motor cortex and from the muscles movements are independent noise, yet it is natural to treat them as noise over which to marginalize the posterior probability density.

In some cases we would also like to treat deficiencies in the computational model as noise, as the following example indicates.

EXAMPLE 7.2: Consider a discrete linear model,

$$Y = AX + E,$$

where $X \in \mathbb{R}^n$ and $E \in \mathbb{R}^m$ are mutually independent Gaussian random variables. Assume that, for some reason, we decide to reduce the model by using the Principal Component Analysis introduced in the previous chapter. We then write

$$X = V_0 \underbrace{(V_0^T X)}_{=X_0} + V_1 \underbrace{(V_1^T X)}_{=X_1} = V_0 X_0 + V_1 X_1,$$

where

$$V = \begin{bmatrix} V_0 & V_1 \end{bmatrix}, \quad V_0 \in \mathbb{R}^{n \times k}, \ V_1 \in \mathbb{R}^{n \times (n-k)}$$

is a partitioning of an orthogonal matrix $V \in \mathbb{R}^{n \times n}$, and we obtain the reduced model

$$Y = \underbrace{AV_0}_{=A_0} X_0 + \underbrace{AV_1 X_1 + E}_{=\widetilde{E}} = A_0 X_0 + \widetilde{E}.$$

In the traditional PCA model reduction, the columns of the orthogonal matrix V are the eigenvectors of the prior covariance matrix. In that case, X_0 and X_1 are independent and, consequently, X_0 and $\widetilde{E}$ are independent. If, however, the matrix V is a sample based estimate of the covariance matrix, the columns of V may not be the eigenvectors of the prior covariance, and the independence is no longer guaranteed.

The point of these two examples is that the likelihood density is not always obtained from the observation model and the noise density alone, but sometimes the prior density needs to be taken into account as well[1].

The central question of this chapter can be stated as follows: assuming that X is a multivariate Gaussian random variable, and some of its components are fixed, what is the distribution of the remaining components? Answering this question will help constructing the likelihood in situations like those described in the examples above, as well as provide other means to analyze posterior densities.

7.1 Gaussian conditional densities

Let $X \in \mathbb{R}^n$ be a Gaussian random variable, and assume for simplicity that its first k components are unknown, $0 < k < n$, while the remaining ones are fixed to known values. To write the density of the k first components conditioned

---

[1] This is yet another instance where the claim that Bayesian statistics equals frequentist statistics plus prior is simply incorrect.

on the knowledge of the last $n - k$ ones, we start with a partitioning of X of the form

$$X = \begin{bmatrix} X_1 \\ X_2 \end{bmatrix} \begin{matrix} \in \mathbb{R}^k \\ \in \mathbb{R}^{n-k} \end{matrix},$$

and we express the probability density of X as the joint density of X_1 and X_2,

$$\pi(x) = \pi(x_1, x_2).$$

The probability density of x_1, under the condition that x_2 is known, is given by

$$\pi(x_1 \mid x_2) \propto \pi(x_1, x_2), \quad x_2 = x_{2,\text{known}}.$$

Observe that if we are investigating a Gaussian linear model with additive noise, and the model is $X_1 = AX_2 + E$, the conditional density above gives the likelihood. If instead we write $X_2 = AX_1 + E$, the conditional density is the posterior density.

EXAMPLE 7.3: Let us consider the simplest possible case where $n = 2$, and assume that the real valued random variables X and Y are zero mean Gaussian variables with symmetric positive definite joint covariance matrix $\Gamma \in \mathbb{R}^{2 \times 2}$,

$$\Gamma = \begin{bmatrix} a & b \\ b & c \end{bmatrix}, \tag{7.1}$$

where

$$a = \mathrm{E}\{X^2\}, \quad b = \mathrm{E}\{XY\}, \quad c = \mathrm{E}\{Y^2\}.$$

The joint probability distribution is then

$$\pi(x, y) \propto \exp\left(-\frac{1}{2} \begin{bmatrix} x & y \end{bmatrix} \Gamma^{-1} \begin{bmatrix} x \\ y \end{bmatrix}\right).$$

We want to find the explicit form of the conditional density $\pi(x \mid y)$ in terms of the components of Γ. Before expressing Γ^{-1} in terms of the components of Γ, let us write

$$\Gamma^{-1} = B = \begin{bmatrix} p & r \\ r & q \end{bmatrix},$$

where the symmetry of B follows from the symmetry of Γ. With this notation, the exponent of the joint density becomes

$$\begin{bmatrix} x & y \end{bmatrix} \Gamma^{-1} \begin{bmatrix} x \\ y \end{bmatrix} = px^2 + 2rxy + qy^2,$$

and after completing the square,

$$px^2 + 2rxy + qy^2 = p\left(x + \frac{ry}{p}\right)^2 + \left(q - \frac{r^2}{p}\right)y^2,$$

we have

$$\pi(x \mid y) \propto \exp\left(-\frac{1}{2}(px^2 + 2rxy + qy^2)\right)$$

$$= \exp\left(-\frac{1}{2}p\left(x + \frac{ry}{p}\right)^2 - \underbrace{\frac{1}{2}\left(q - \frac{r^2}{p}\right)y^2}_{\text{independent of } x}\right)$$

$$\propto \exp\left(-\frac{1}{2}p\left(x + \frac{ry}{p}\right)^2\right),$$

since the term independent of x contributes only in the form of a normalizing constant. We conclude that the probability density of x conditioned on y is a Gaussian density with mean and variance

$$\bar{x} = -\frac{ry}{p}, \quad \sigma^2 = \frac{1}{p},$$

respectively. We are now ready to take a second look at Γ^{-1}, since, after all, we may want to express the quantities p, q and r in terms of the components of Γ.

Observe that since Γ is positive definite, that is,

$$v^T \Gamma v > 0 \quad \text{for all } v \neq 0,$$

by choosing $v = e_1 = \begin{bmatrix} 1 & 0 \end{bmatrix}^T$ or $v = e_2 = \begin{bmatrix} 0 & 1 \end{bmatrix}^T$, we have

$$a = e_1^T \Gamma e_1 > 0, \quad c = e_2^T \Gamma e_2 > 0.$$

Let us now return to Γ^{-1}. It is elementary to verify that

$$\Gamma^{-1} = \frac{1}{ac - b^2}\begin{bmatrix} c & -b \\ -b & a \end{bmatrix},$$

where $ac - b^2 = \det(\Gamma) \neq 0$. Expressing the determinant of Γ in the form

$$ac - b^2 = a\underbrace{\left(c - \frac{b^2}{a}\right)}_{=\tilde{a}} = c\underbrace{\left(a - \frac{b^2}{c}\right)}_{=\tilde{c}},$$

we notice that, if

$$\tilde{a} = c - \frac{b^2}{a} \neq 0, \quad \tilde{c} = a - \frac{b^2}{c} \neq 0,$$

we can also express it in the form

$$\det(\Gamma) = a\tilde{a} = c\tilde{c}.$$

We now write the inverse of Γ in terms of these newly defined quantities. Since

$$\frac{a}{\det(\Gamma)} = \frac{1}{\tilde{a}}, \quad \frac{c}{\det(\Gamma)} = \frac{1}{\tilde{c}},$$

and

$$\frac{b}{\det(\Gamma)} = \frac{b}{a\tilde{a}} = \frac{b}{c\tilde{c}},$$

we have that

$$\Gamma^{-1} = \begin{bmatrix} 1/\tilde{a} & -b/(a\tilde{a}) \\ -b/(c\tilde{c}) & 1/\tilde{c} \end{bmatrix} = \begin{bmatrix} p & r \\ r & q \end{bmatrix}. \tag{7.2}$$

Note that from (7.2) we have two alternative expressions for r,

$$r = -\frac{b}{a\tilde{a}} = -\frac{b}{c\tilde{c}}.$$

Clearly the key players in calculating the inverse of Γ are the quantities $\tilde{a}$ and $\tilde{c}$, which are called the *Schur complements* of a and c, respectively. In terms of the Schur complements, the conditional density can be written as

$$\pi(x \mid y) \propto \exp\left(-\frac{1}{2\sigma^2}(x - \bar{x})^2\right),$$

with mean

$$\bar{x} = -\frac{r}{p}y = \frac{b}{\tilde{a}a}\tilde{a}y = \frac{b}{a}y,$$

and variance

$$\operatorname{var}(x) = \sigma^2 = \frac{1}{p} = \tilde{a},$$

respectively.

We are now ready to consider the general multivariate case, trying to mimic what we did in the simple case. Assuming that $x \in \mathbb{R}^n$ is a zero mean Gaussian random variable with covariance matrix $\Gamma \in \mathbb{R}^{n\times n}$, the joint probability density of $x_1 \in \mathbb{R}^k$ and $x_2 \in \mathbb{R}^{n-k}$ is

$$\pi(x_1, x_2) \propto \exp\left(-\frac{1}{2}x^{\mathrm{T}}\Gamma^{-1}x\right). \tag{7.3}$$

To investigate how this expression depends on x_1 when x_2 is fixed, we start by partitioning the covariance matrix

$$\Gamma = \begin{bmatrix} \Gamma_{11} & \Gamma_{12} \\ \Gamma_{21} & \Gamma_{22} \end{bmatrix} \in \mathbb{R}^{n\times n}, \tag{7.4}$$

where

$$\Gamma_{11} \in \mathbb{R}^{k \times k}, \quad \Gamma_{22} \in \mathbb{R}^{(n-k) \times (n-k)}, \quad k < n,$$

and

$$\Gamma_{12} = \Gamma_{21}^{\mathrm{T}} \in \mathbb{R}^{k \times (n-k)}.$$

Proceeding as in the two dimensional example, we denote Γ^{-1} by B, and we partition it according to the partition of Γ,

$$\Gamma^{-1} = B = \begin{bmatrix} B_{11} & B_{12} \\ B_{21} & B_{22} \end{bmatrix} \in \mathbb{R}^{n \times n}. \tag{7.5}$$

We then write the quadratic form appearing in the exponential of (7.3) as

$$x^{\mathrm{T}} B x = x_1^{\mathrm{T}} B_{11} x_1 + 2 x_1^{\mathrm{T}} B_{12} x_2 + x_2^{\mathrm{T}} B_{22} x_2 \tag{7.6}$$

$$= \left(x_1 + B_{11}^{-1} B_{12} x_2 \right)^{\mathrm{T}} B_{11} \left(x_1 + B_{11}^{-1} B_{12} x_2 \right) + \underbrace{x_2^{\mathrm{T}} \left(B_{22} - B_{21} B_{11}^{-1} B_{12} \right) x_2}_{\text{independent of } x_1}.$$

This is the key equation when considering conditional densities. Observing that the term in (7.6) that does not depend on x_1 contributes only to a multiplicative constant, we have

$$\pi(x_1 \mid x_2) \propto \exp\left(-\frac{1}{2} \left(x_1 + B_{11}^{-1} B_{12} x_2 \right)^{\mathrm{T}} B_{11} \left(x_1 + B_{11}^{-1} B_{12} x_2 \right) \right).$$

We therefore conclude that the conditional density is Gaussian,

$$\pi(x_1 \mid x_2) \propto \exp\left(-\frac{1}{2} (x_1 - \bar{x}_1)^{\mathrm{T}} C^{-1} (x_1 - \bar{x}_1) \right),$$

with mean

$$\bar{x}_1 = -B_{11}^{-1} B_{12} x_2$$

and covariance matrix

$$C = B_{11}^{-1}.$$

To express these quantities in terms of the covariance matrix Γ, we need to introduce multivariate Schur complements. For this purpose we now derive a formula similar to (7.2) for general $n \times n$ matrices. Let us consider a symmetric positive definite matrix $\Gamma \in \mathbb{R}^{n \times n}$, partitioned according to (7.5).

Since Γ is positive definite, both Γ_{11} and Γ_{22} are also. In fact, for any $x_1 \in \mathbb{R}^k$, $x_1 \neq 0$,

$$x_1^{\mathrm{T}} \Gamma_{11} x_1 = \begin{bmatrix} x_1^{\mathrm{T}} & 0 \end{bmatrix} \begin{bmatrix} \Gamma_{11} & \Gamma_{12} \\ \Gamma_{21} & \Gamma_{22} \end{bmatrix} \begin{bmatrix} x_1 \\ 0 \end{bmatrix} > 0,$$

showing the positive definiteness of Γ_{11}. The proof that Γ_{22} is positive definite is analogous.

To compute Γ^{-1} using the block partitioning, we solve the linear system

$$\Gamma x = y$$

in block form, that is, we partition x and y as

$$x = \begin{bmatrix} x_1 \\ x_2 \end{bmatrix} \begin{matrix} \in \mathbb{R}^k \\ \in \mathbb{R}^{n-k} \end{matrix}, \quad y = \begin{bmatrix} y_1 \\ y_2 \end{bmatrix} \begin{matrix} \in \mathbb{R}^k \\ \in \mathbb{R}^{n-k} \end{matrix},$$

and write

$$\Gamma_{11}x_1 + \Gamma_{12}x_2 = y_1, \tag{7.7}$$
$$\Gamma_{21}x_1 + \Gamma_{22}x_2 = y_2. \tag{7.8}$$

Solving the second equation for x_2, which can be done because Γ_{22}, being positive definite, is also invertible, we have

$$x_2 = \Gamma_{22}^{-1}(y_2 - \Gamma_{21}x_1),$$

and substituting it into the first equation, after rearranging the terms, yields

$$(\Gamma_{11} - \Gamma_{12}\Gamma_{22}^{-1}\Gamma_{21})x_1 = y_1 - \Gamma_{12}\Gamma_{22}^{-1}y_2.$$

We define the Schur complement of Γ_{22} to be

$$\widetilde{\Gamma}_{22} = \Gamma_{11} - \Gamma_{12}\Gamma_{22}^{-1}\Gamma_{21}.$$

With this notation,

$$x_1 = \widetilde{\Gamma}_{22}^{-1}y_1 - \widetilde{\Gamma}_{22}^{-1}\Gamma_{12}\Gamma_{22}^{-1}y_2. \tag{7.9}$$

Similarly, by solving (7.7) for x_1 first and plugging it into (7.8), we may express x_2 as

$$x_2 = \widetilde{\Gamma}_{11}^{-1}y_2 - \widetilde{\Gamma}_{11}^{-1}\Gamma_{21}\Gamma_{11}^{-1}y_1 \tag{7.10}$$

in terms of the Schur complement of Γ_{11},

$$\widetilde{\Gamma}_{11} = \Gamma_{22} - \Gamma_{21}\Gamma_{11}^{-1}\Gamma_{12}.$$

Collecting (7.9) and (7.10) into

$$\begin{bmatrix} x_1 \\ x_2 \end{bmatrix} = \begin{bmatrix} \widetilde{\Gamma}_{22}^{-1} & -\widetilde{\Gamma}_{22}^{-1}\Gamma_{12}\Gamma_{22}^{-1} \\ -\widetilde{\Gamma}_{11}^{-1}\Gamma_{21}\Gamma_{11}^{-1} & \widetilde{\Gamma}_{11}^{-1} \end{bmatrix} \begin{bmatrix} y_1 \\ y_2 \end{bmatrix},$$

we deduce that

$$\Gamma^{-1} = \begin{bmatrix} \widetilde{\Gamma}_{22}^{-1} & -\widetilde{\Gamma}_{22}^{-1}\Gamma_{12}\Gamma_{22}^{-1} \\ -\widetilde{\Gamma}_{11}^{-1}\Gamma_{21}\Gamma_{11}^{-1} & \widetilde{\Gamma}_{11}^{-1} \end{bmatrix} = \begin{bmatrix} B_{11} & B_{12} \\ B_{21} & B_{22} \end{bmatrix},$$

which is the formula that we were looking for.

In summary, we have derived the following result: *the conditional density* $\pi(x_1 \mid x_2)$ *is a Gaussian probability distribution whose conditional mean (CM) and conditional covariance are*

$$\bar{x}_1 = -B_{11}^{-1} B_{12} x_2 = \Gamma_{12} \Gamma_{22}^{-1} x_2,$$

and

$$C = B_{11}^{-1} = \tilde{\Gamma}_{22},$$

respectively.

Now we present some examples of how to use of this result.

7.2 Interpolation, splines and conditional densities

As an application of the computation of conditional densities for Gaussian distributions, and also as an introduction to other topics that will be discussed later, consider the following classical interpolation problem:

Find a smooth function f defined on an interval $[0, T]$, *that satisfies the constraints*

$$f(t_j) = b_j \pm e_j, \quad 1 \le j \le m,$$

where $0 = t_1 < t_2 < \cdots < t_m = T$. We refer to the b_j as the data and to the e_j as error bounds[2].

Since in the statement of the problem no exact definition of smoothness or of the error bounds are given, we will interpret them later subjectively.

A classical solution to this problem uses the cubic splines, i.e., curves that between the interpolation points are third order polynomials, glued together at the observation points so that the resulting piecewise defined function and its derivatives, up to the second order, are continuous. The Bayesian solution that we are interested in is based on conditioning and, it turns out, can be orchestrated so that the realization with highest posterior probability corresponds to a spline.

As a prelude, consider a simplified version of the problem, which assumes that only the values at the endpoints are given, and that they are known *exactly*,

$$f(0) = b_1, \quad f(T) = b_2. \tag{7.11}$$

An obvious solution to this interpolation problem is a linear function passing through the data points. In the following construction, this trivial solution will be shown to be also the most probable realization with respect to the prior that we introduce. As it turns out, in this simplified case the use of Schur complements can be avoided, while they become useful when the data is collected also in interior points.

---

[2] This expression is borrowed from the engineering literature, where it is common to refer to measurements as mean ± standard deviation to mean that an error with the indicated standard deviation is expected in the measured data.

We start by discretizing the problem: divide the support of the function into n equal subintervals, and let

$$x_j = f(s_j), \quad s_j = jh, \quad h = \frac{T}{n}, \quad 0 \le j \le n.$$

The smoothness of the solution is favored by imposing a prior condition that the values of the function at interior points are not very different from the average of the adjacent values, that is,

$$x_j \approx \frac{1}{2}(x_{j-1} + x_{j+1}), \quad 1 \le j \le n-1.$$

Similarly to the construction in Example 3.8, we write a stochastic Gaussian prior model for the random variable $X \in \mathbb{R}^n$ of the form

$$X_j = \frac{1}{2}(X_{j-1} + X_{j+1}) + W_j, \quad 1 \le j \le n-1,$$

and the innovation vector $W \in \mathbb{R}^{n-1}$ with components W_j is modelled as a multivariate Gaussian white noise process,

$$W \sim \mathcal{N}(0, \gamma^2 I).$$

The variance γ^2 is a measure of how much we believe in the averaging model. Notice that forcing the mean value assumption with no innovation term would give the linear interpolation solution.

In vector form, the prior model can be written as

$$LX = W,$$

where L is the second order finite difference matrix,

$$L = \frac{1}{2} \begin{bmatrix} -1 & 2 & -1 & & \\ & -1 & 2 & -1 & \\ & & \ddots & \ddots & \ddots \\ & & & -1 & 2 & -1 \end{bmatrix} \in \mathbb{R}^{(n-1) \times (n+1)}. \tag{7.12}$$

To discuss the conditional density, let us partition the entries of x in the two vectors

$$x' = \begin{bmatrix} x_1 \\ x_2 \\ \vdots \\ x_{n-1} \end{bmatrix} \in \mathbb{R}^{n-1}, \quad x'' = \begin{bmatrix} x_0 \\ x_n \end{bmatrix} \in \mathbb{R}^2.$$

The probability density corresponding to the prior model above is then

$$\pi(x' \mid x'') \propto \pi(x', x'') = \pi(x) \propto \exp\left(-\frac{1}{2\gamma^2} \|Lx\|^2\right),$$

where

$$x'' = x''_{known} = \begin{bmatrix} b_1 \\ b_2 \end{bmatrix} = b.$$

Thus, the density defines the probability distribution of the interior points assuming that the boundary values are given, and since its maximum occurs at the solution of the equation $Lx = 0$, implying that x is a discretization of a linear function, it favors solutions with small oscillations.

We now apply the conditioning formulas and write the conditional density explicitly in terms of x'. Assume that by permuting the entries of x, the two observed values are the last two components of x, write x in the form

$$x = \begin{bmatrix} x' \\ x'' \end{bmatrix}.$$

and partition the matrix L accordingly, after having permuted the columns to follow the ordering of the components of x,

$$L = \begin{bmatrix} L_1 & L_2 \end{bmatrix}, \quad L_1 \in \mathbb{R}^{(n-1) \times (n-1)}, \quad L_2 \in \mathbb{R}^{(n-1) \times 2}.$$

It can be checked that the matrix L_1 is invertible. Since

$$Lx = L_1 x' + L_2 x'' = L_1(x' + L_1^{-1} L_2 x''),$$

it follows that the conditional probability density

$$\pi(x' \mid x'') \propto \exp\left(-\frac{1}{2\gamma^2} \|L_1(x' + L_1^{-1} L_2 x'')\|^2 \right)$$

$$= \exp\left(-\frac{1}{2}(x' + L_1^{-1} L_2 x'')^{\mathrm{T}} \left(\frac{1}{\gamma^2} L_1^{\mathrm{T}} L_1 \right) (x' + L_1^{-1} L_2 x'') \right),$$

is Gaussian,

$$\pi(x' \mid x'' = b) \sim \mathcal{N}(\overline{x}', \Gamma'),$$

with mean and covariance

$$\overline{x}' = -L_1^{-1} L_2 b, \quad \Gamma' = \gamma^2 (L_1^{\mathrm{T}} L_1)^{-1},$$

respectively.

Before considering the case where we have more data points, assume that, instead of the exact data (7.11), we know the data up to an additive error. We interpret the error bar condition in the problem setup as stating that the additive error is Gaussian

$$X_0 = b_1 + E_1, \quad X_n = b_2 + E_2, \quad E_j \sim \mathcal{N}(0, \sigma_j^2),$$

which implies that the marginal density of $X'' = [X_1 \ X_n]^{\mathrm{T}}$ is

$$\pi(x'') \propto \exp\left(-\frac{1}{2}(x'' - b)^{\mathrm{T}} C^{-1}(x'' - b)\right),$$

where

$$C = \begin{bmatrix} \sigma_1^2 & \\ & \sigma_2^2 \end{bmatrix}.$$

Since now the determination of both x' *and* x'' from the data is of concern – x'' is no longer known exactly – we write the joint probability density as

$$\pi(x) = \pi(x', x'') = \pi(x' \mid x'')\pi(x''),$$

or, explicitly,

$$\pi(x) \propto \exp\left(-\frac{1}{2\gamma^2}\|L_1(x' + L_1^{-1}L_2 x'')\|^2 - \frac{1}{2}(x'' - b)^{\mathrm{T}} C^{-1}(x'' - b)\right).$$

To better understand the significance of this formula, some matrix manipulations are required. We simplify the notations by assuming that $\gamma = 1$, a condition which is easy to satisfy simply by scaling L by $1/\gamma$.

Let's begin by writing

$$\|L_1(x' + L_1^{-1}L_2 x'')\|^2 = \|Lx\|^2 = \left\|\begin{bmatrix} L_1 & L_2 \end{bmatrix} \begin{bmatrix} x' \\ x'' \end{bmatrix}\right\|^2,$$

and factorizing the matrix C^{-1} as

$$C^{-1} = K^{\mathrm{T}}K, \quad K = \begin{bmatrix} 1/\sigma_1 & \\ & 1/\sigma_2 \end{bmatrix},$$

so that

$$(x'' - b)^{\mathrm{T}} C^{-1}(x'' - b) = \|K(x'' - b)\|^2$$

$$= \left\|\begin{bmatrix} 0 & K \end{bmatrix} \begin{bmatrix} x' \\ x'' \end{bmatrix} - Kb\right\|^2.$$

We can now write the probability density of X as

$$\pi(x) \propto \exp\left(-\frac{1}{2}\left\|\begin{bmatrix} L_1 & L_2 \end{bmatrix} \begin{bmatrix} x' \\ x'' \end{bmatrix}\right\|^2 - \frac{1}{2}\left\|\begin{bmatrix} 0 & K \end{bmatrix} \begin{bmatrix} x' \\ x'' \end{bmatrix} - Kb\right\|^2\right)$$

$$= \exp\left(-\frac{1}{2}\left\|\begin{bmatrix} L_1 & L_2 \\ 0 & K \end{bmatrix} \begin{bmatrix} x' \\ x'' \end{bmatrix} - \begin{bmatrix} 0 \\ Kb \end{bmatrix}\right\|^2\right)$$

$$= \exp\left(-\frac{1}{2}\|\widetilde{L}(x - \overline{x})\|^2\right),$$

where

$$\widetilde{L} = \begin{bmatrix} L_1 & L_2 \\ 0 & K \end{bmatrix},$$

and

$$\overline{x} = \widetilde{L}^{-1} \begin{bmatrix} 0 \\ Kb \end{bmatrix}. \tag{7.13}$$

We therefore conclude that the probability density of X is Gaussian

$$\pi(x) \sim \mathcal{N}(\overline{x}, \Gamma),$$

with the mean $\overline{x}$ from (7.13) and covariance

$$\Gamma = (\widetilde{L}^{\mathrm{T}}\widetilde{L})^{-1}.$$

We return now to the original interpolation problem and add data points inside the interval $[0, T]$. For simplicity, let us assume that the data points t_j coincide with discretization points s_k so that, in the case of exact data,

$$x_{k_j} = f(t_j) = b_j, \quad 1 \le j \le m. \tag{7.14}$$

Partition x as

$$x = \begin{bmatrix} x' \\ x'' \end{bmatrix}, \quad x'' = \begin{bmatrix} x_{k_1} \\ \vdots \\ x_{k_m} \end{bmatrix} = \begin{bmatrix} b_1 \\ \vdots \\ b_m \end{bmatrix},$$

permuting the elements of x so that the components corresponding to the data are the last ones.

Among the several possible ways to modify the procedure discussed above so that it can be applied to more general problems, we choose one which extends easily to higher dimensions. Let $L \in \mathbb{R}^{(n-1)\times(n+1)}$ be the matrix (7.12). After permuting the columns of L to correspond to the ordering of the elements of x, partition L according to the partition of x,

$$L = \begin{bmatrix} L_1 & L_2 \end{bmatrix}, \quad L_1 \in \mathbb{R}^{(n-1)\times(n-m+1)}, \quad L_2 \in \mathbb{R}^{(n-1)\times m}.$$

Observe that since L_1 is no longer a square matrix, it is obviously not invertible and we cannot proceed as in the simple case when the data was given at the endpoints only. Therefore let

$$B = L^{\mathrm{T}}L = \begin{bmatrix} B_{11} & B_{12} \\ B_{21} & B_{22} \end{bmatrix},$$

where

$$B_{11} = L_1^{\mathrm{T}}L_1 \in \mathbb{R}^{(n-m+1)\times(n-m+1)}, \quad B_{22} = L_2^{\mathrm{T}}L_2 \in \mathbb{R}^{m\times m},$$

and

$$B_{12} = L_1^{\mathrm{T}}L_2 = B_{21}^{\mathrm{T}} \in \mathbb{R}^{(n-m+1)\times m}.$$

Similarly as in the previous section, the conditional density can be written in the form

$$\pi(x' \mid x'' = b) \propto \exp\left(-\frac{1}{2}(x' + B_{11}^{-1}B_{12}b)^{\mathrm{T}}B_{11}(x' + B_{11}^{-1}B_{12}b)\right).$$

If the observations contain additive Gaussian error,

$$X_{k_j} = b_j + E_j, \quad E_j \sim \mathcal{N}(0, \sigma_j^2),$$

we introduce the covariance matrix

$$C = \begin{bmatrix} \sigma_1^2 & & \\ & \ddots & \\ & & \sigma_m^2 \end{bmatrix},$$

and write

$$\pi(x'') \propto \exp\left(-\frac{1}{2}(x'' - b)^{\mathrm{T}}C^{-1}(x'' - b)\right),$$

leading to a joint probability density of the form

$$\pi(x) = \pi(x' \mid x'')\pi(x'')$$

$$\propto \exp\left(-\frac{1}{2}(x' + B_{11}^{-1}B_{12}x'')^{\mathrm{T}}B_{11}(x' + B_{11}^{-1}B_{12}x'')\right.$$

$$\left. -\frac{1}{2}(x'' - b)^{\mathrm{T}}C^{-1}(x'' - b)\right).$$

To extract the covariance and mean from this expression, the terms need to be rearranged. After a straightforward but tedious exercise in completing the square, we find that

$$\pi(x) \propto \exp\left(\frac{1}{2}(x - \bar{x})^{\mathrm{T}}\Gamma^{-1}(x - \bar{x})\right),$$

where

$$\Gamma = \begin{bmatrix} B_{11} & B_{12} \\ B_{21} & B_{21}B_{11}^{-1}B_{12} + C^{-1} \end{bmatrix}^{-1}, \tag{7.15}$$

and

$$\bar{x} = \Gamma \begin{bmatrix} 0 \\ b \end{bmatrix}. \tag{7.16}$$

The mean $\bar{x}$ is a good candidate for a single solution curve to the interpolation problem.

Consider now a computed example where the formulas derived above are applied.

EXAMPLE 7.4: The support of the function is a unit interval, i.e., $T = 1$, and the number of data points is $m = 6$. We divide the interval into $n = 60$ equal subintervals. To fix the parameter γ, recall that it conveys our uncertainty of how well the function obeys the mean value rule. Thus, if we assume that a typical increment $|x_j - x_{j-1}|$ over a subinterval is of order $1/n$, we may be confident that the prior is not too committal if γ is of the same order of magnitude. Therefore in our computed example, we set $\gamma = 1/n$.

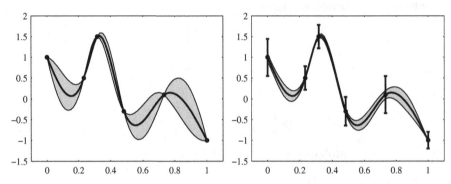

Fig. 7.1. Mean and credibility envelopes of two standard deviations with noiseless data (left) and with noisy data (right), the error bars indicating an uncertainty of two standard deviations.

Given the values b_j, we calculate the mean and covariance of x in the two cases when the values are exactly known, and where they are normally distributed with variance σ_j^2.

In Figure 7.1, we have plotted the solution of the interpolation problem when the data is exact in the left panel, and normally distributed in the right panel, where the error bars of two standard deviations are also indicated. In both cases, we have plotted the mean curve and a *pointwise predictive output envelope* corresponding to two standard deviations, which is the shaded area around the mean curve is bounded by the curves

$$s_j \mapsto \overline{x}_j \pm 2\Gamma_{jj}^{1/2},$$

with the understanding that in the case of exact data, at data points $\overline{x}_j$ coincides with the data and the variance vanishes. Comparing the two plots in Figure 7.1, we notice that the mean curves with exact and noisy data are similar while the predictive envelopes are rather different. Clearly, when forcing the solutions to pass through the data points, the uncertainty in the intervals between data points increases significantly.

The importance of taking into consideration the information carried by the predictive envelopes become even more clear when we consider a realistic interpolation problem. A typical example is the problem of interpolating geographic temperature distributions. Starting from a sparse network of observation stations where the temperature is measured, we interpolate the temperature in the surrounding domain. The conditioning approach derived above in the one dimensional case extends in a natural way to higher dimensions.

EXAMPLE 7.5: Consider the interpolation problem in two dimensions where we want to find a smooth surface with a square support that passes through a given set of points in $\mathbb{R}^3$. Assume that the domain is discretized into $N = n \times n$ pixels, and $X \in \mathbb{R}^N$ is a random vector representing the surface as a pixelized image. We assume that some of the pixel values are fixed,

$$X_j = b_j, \quad j \in I_2 \subset I = \{1, 2, \ldots, N\}, \quad I_1 = I \setminus I_2,$$

and we want to find the conditional distribution of the vector of the remaining pixels,

$$X'' = \left[X_j\right]_{j \in I_1},$$

conditioned on

$$X' = \left[X_j\right]_{j \in I_2} = b.$$

We assume a priori that X is smooth and nearly vanishing at the boundaries, encoding this belief in a smoothness prior

$$\pi(x) \propto \exp\left(-\frac{1}{2\gamma^2}\|Lx\|^2\right),$$

where L is the second order finite difference operator constructed in Example 3.8, corresponding to the mask

$$\begin{bmatrix} & -1/4 & \\ -1/4 & 1 & -1/4 \\ & -1/4 & \end{bmatrix}.$$

This prior model conveys the idea that each pixel value is the mean of the four neighboring pixels plus a random white noise innovation, the surface values being extended by zeros outside the image view.

Instead of applying the formulas that have been derived in the one dimensional case, let us look directly the Matlab code used to solve the problem, assuming that the matrix L has already been computed.

```
B = L'*L;

% Conditioning: I2 defines the points that are known,
% I1 points to free pixels

I = reshape([1:n^2],n,n);
I2 = [I(15,15) I(15,25) I(15,40) I(35,25)];
```

```
x2 = [15;-20;10;10];
I1 = setdiff(I(:),I2);

% Conditional covariance of x1

B11 = B(I1,I1);
B12 = B(I1,I2);

% Calculating and plotting the conditional mean value
% surface

x1mean = -B11\(B12*x2);
xplot = zeros(n^2,1);
xplot(I1) = x1mean;
xplot(I2) = x2;
figure(1)
surfl(reshape(xplot,n,n)), shading flat, colormap(gray)

% Generate a few random draws from the distribution

R = chol(B11);     % Whitening matrix

ndraws = 2;
for j = 1:ndraws
    xdraw1 = x1mean + R\randn(length(I1),1);
    xdraw = zeros(n^2,1);
    xdraw(I1) = xdraw1;
    xdraw(I2) = x2;

    figure(1+j)
    imagesc(flipud(reshape(xdraw,n,n))), shading flat
    axis('square'), colormap(gray)
end

% Plotting the standard deviation surface

STDs = zeros(length(I1),1);
for j = 1:length(I1)
    ej = zeros(length(I1),1); ej(j)=1;
    STDs(j) = norm(R'\ej);
end

STDsurf = zeros(n^2,1);
STDsurf(I1) = STDs;
figure(ndraws + 2)
```

```
surfl(reshape(varsurf,n,n))
shading flat, colormap(gray)
```

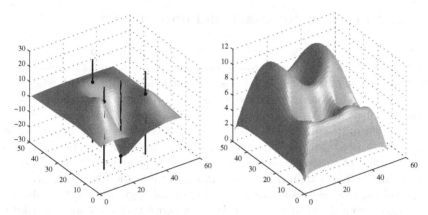

Fig. 7.2. The conditional mean surface (left) with the fixed values marked by dots. The surface plotted in the right panel shows the square root of the diagonal entries of the conditional covariance matrix, which gives the standard deviation of each pixel. Notice that the standard deviation of the fixed pixels is, of course, zero. Also notice the effect of continuation by zero beyond the boundaries.

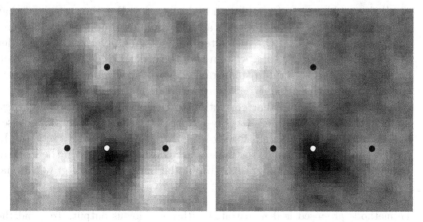

Fig. 7.3. Two random draws from the conditional density. The data points are marked by dots.

The mean and variance surfaces are shown in Figure 7.2, and two random draws are displayed in Figure 7.3. The standard deviation surface plot gives an indication of how much credibility the predictions of the temperature between the observation points have, based on the data and the prior model. In Figure 7.3, two random draws from the conditional density are displayed as gray scale images.

7.3 Envelopes, white swans and dark matter

In the previous section, we pointed out that the Bayesian analysis of the interpolation problem gave not only a probable solution, but also credibility envelopes around it. These envelopes are particularly important when the models are used to make predictions about the behavior of quantities that cannot be observed. An isolated predicted value, curve or surface can be useless, or even misleading unless we have some way of assessing how much it should be trusted[3]. In the history of medicine, for instance, there are several examples of catastrophic results as the consequence of single predictive outputs made from carefully collected and analyzed experimental data.

In this section we return to the philosophical argument that producing one, two or even three white swans does not prove that all swans are white. Predictions are always extrapolations: the most interesting predictions are for quantities or setting for which no data are available. Therefore, an obvious choice to demonstrate the importance of predictive envelopes is to see them in action on a simple extrapolation problem.

EXAMPLE 7.6: One of the fundamental beliefs in physics is that the gravitational force exerted by a mass on a test particle is a linear function of the particle's mass. This theory has been tested with objects of relatively small size and it seems to hold with high accuracy. But the statement that it holds universally for all masses is a rather bold extrapolation of local linearity. Considering that the estimated mass of the universe is of the order of 10^{60}, the mass of the solar system is of the order 10^{30}, our experience on gravity is limited to a quite small range. At this point one may wonder if the dilemma of the missing dark matter could not, after all, be explained by adding a small nonlinear correction to the linear gravitational theory. To test this, let's assume that we perform 500 measurements of masses and associated gravitational force of the Earth, (m_j, F_j), $1 \leq j \leq 500$. In our Gedankenexperiment[4], the masses are chosen randomly in the interval between 10^{-3} and 10^4 kilograms. To put the linear theory of gravitational

---

[3] In most of the non-statistical inverse problems literature, the performance of the methods proposed is demonstrated with one dubious output. To avoid the misunderstanding that we say this to blame others, we remark that in numerous articles written by the authors, a wealth of examples adhering to the "dubious output paradigm" can be found!

[4] The German term for thought experiment, that was strongly advocated by Einstein, goes back to the Danish physicist and chemist Hans Christian Ørsted (1777–1851), a keen reader of the philosophy of Immanuel Kant.

force in pole position, we calculate the simulated data using the linear formula

$$F_i = gm_i + e_i,$$

where $g = 9.81$ m/s^2, and draw the error e_i from a Gaussian, normally distributed with standard deviation $\sigma = 10^{-10}$, which means that we assume a tremendous accuracy of the data.

Being slightly doubtful about the validity of the linear theory of gravity, we consider the possibility of enriching it with a possible quadratic correction, to obtain the new model

$$F(m) = x_1 m + x_2 m^2.$$

We express, however, a strong confidence in the validity of the linear model via the prior model

$$\pi_{\mathrm{prior}}(x) \propto \exp\left(-\frac{1}{2\gamma^2}(|x_1 - g|^2 + |x_2|^2)\right), \quad \gamma = 10^{-14},$$

stating that, by any reasonable measure, we are almost certain of the validity of the linear model with gravitational acceleration equal to g.

The question that we now ask is: *Given the strong belief in the linear model and the data, what does this model predict the gravitational force exerted by the mass of the Earth to be for masses in an interval $[m_{\min}, m_{\max}]$ which is beyond the observable and measurable range?*

To answer this question we write a likelihood model, admitting that the measurements may contain a negligible Gaussian error of standard deviation σ given above. Letting

$$A = \begin{bmatrix} m_1 & m_1^2 \\ m_2 & m_2^2 \\ \vdots & \vdots \\ m_n & m_n^2 \end{bmatrix}, F = \begin{bmatrix} F_1 \\ F_2 \\ \vdots \\ F_n \end{bmatrix},$$

the likelihood is of the form

$$\pi(F \mid x) \propto \exp\left(-\frac{1}{2\sigma^2}\|Ax - F\|^2\right),$$

leading to the posterior density

$$\pi(x \mid F) \propto \exp\left(-\frac{1}{2\sigma^2}\|Ax - F\|^2 - \frac{1}{2\gamma^2}\|x - \overline{x}\|^2\right)$$

$$= \exp\left(-\frac{1}{2\sigma^2}\left\|\begin{bmatrix} A \\ \delta I \end{bmatrix} x_{\mathrm{c}} - \begin{bmatrix} F_{\mathrm{c}} \\ 0 \end{bmatrix}\right\|^2\right) \quad \delta = \frac{\sigma}{\gamma},$$

where

$$\bar{x} = \begin{bmatrix} g \\ 0 \end{bmatrix}, \quad x_c = x - \bar{x}, \quad F_c = F - A\bar{x}.$$

Since we are interested in the predictions of this model, we generate a large sample of vectors x^j, $1 \leq j \leq N$ and the corresponding sample of prediction curves,

$$m \mapsto x_1^j m + x_2^j m^2, \quad m_{min} \leq m \leq m_{max}.$$

Then, for each value of m we calculate intervals that contains 90% of the values $x_1^j m + x_2^j m^2$, $1 \leq j \leq N$. These intervals define the 90% pointwise predictive output envelope of the model.

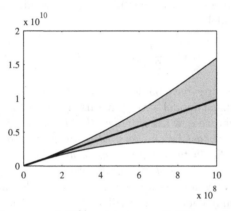

Fig. 7.4. Predicted mean and 90% pointwise predictive envelope of extrapolation problem.

The curve corresponding to the posterior mean together with the predictive output envelope is plotted in Figure 7.4. The number of sample points in this example is $N = 1000$, and the interval of extrapolation is $[1, 10^9]$. Despite the strong prior belief in the linear model and the good quality of the data – which should have favored the linear model that was used to generate it – the belief envelope is amazingly wide, and orders of magnitude away from mean extrapolation.

Although the model is not taking into account the evidence we might have of the linearity of the gravity, at this point one might wonder if the dark matter is but a white swan.

The themes discussed in this chapter will be further developed in the next chapter and the exercises are postponed to the end of that chapter.

8

More applications of the Gaussian conditioning

In the previous chapter we developed tools to calculate the Gaussian conditional probability densities using partitioning of the joint covariance matrix and Schur complements. In this chapter we apply these tools to a few more complex examples. We begin with illustrating how to analyze linear Gaussian models with additive noise, a procedure with a wide range of applicability, and which will serve as a basis to treat more general cases.

8.1 Linear inverse problems

Assume that we have a linear model,

$$B = AX + E,$$

where the pair (X, E) is Gaussian. Although it is rather common to assume that X and E are mutually independent, in general this is neither necessary nor easily justified, as we pointed out in the previous chapter. The question that we address here is the calculation of the conditional density $\pi(x \mid y)$. We recall that when X and E are mutually independent, the Bayes formula states that

$$\pi(x \mid y) \propto \pi_{\text{prior}}(x)\pi_{\text{noise}}(b - Ax).$$

Assuming that $X \sim \mathcal{N}(x_0, \Gamma)$, $E \sim \mathcal{N}(e_0, \Sigma)$, and that the inverses of the covariance matrices have symmetric factorizations,

$$\Gamma^{-1} = L^{\mathrm{T}} L, \quad \Sigma^{-1} = S^{\mathrm{T}} S,$$

we can write the posterior density in the form

$$\pi(x \mid y) \propto \exp\left(-\frac{1}{2}\|L(x - x_0)\|^2 - \frac{1}{2}\|S(b - e_0 - Ax)\|^2 \right).$$

This useful expression for the posterior density is not the most general one. As it was pointed out in Examples 7.1 and 7.2, the mutual independence of the unknown X and of the noise E is a special condition that cannot always be assumed.

To derive a more general form, let us assume for simplicity that X and E have zero mean. If this is not the case, we can always consider instead the variables $X - x_0$ and $E - e_0$, where x_0 and e_0 are the means. With this assumption, B has also mean zero:

$$E\{B\} = A E\{X\} + E\{E\} = 0.$$

Let us define the covariance matrices,

$$\Gamma_{xx} = E\{XX^T\} \in \mathbb{R}^{n \times n},$$

$$\Gamma_{xb} = \Gamma_{bx}^T = E\{XB^T\} \in \mathbb{R}^{n \times m}, \tag{8.1}$$

$$\Gamma_{bb} = E\{BB^T\} \in \mathbb{R}^{m \times m},$$

and the joint covariance matrix,

$$E\left\{ \begin{bmatrix} X \\ B \end{bmatrix} [X^T \ B^T] \right\} = \begin{bmatrix} \Gamma_{xx} & \Gamma_{xb} \\ \Gamma_{bx} & \Gamma_{bb} \end{bmatrix} \in \mathbb{R}^{(n+m) \times (n+m)}.$$

To calculate the posterior covariance, we note that it was shown earlier that the conditional density $\pi(x \mid y)$ is a Gaussian probability distribution with center point

$$x_{\mathrm{CM}} = \Gamma_{xb} \Gamma_{bb}^{-1} b, \tag{8.2}$$

and with covariance matrix the Schur complement of Γ_{bb}.

$$\Gamma_{\mathrm{post}} = \widetilde{\Gamma}_{bb}. \tag{8.3}$$

To calculate the covariances (8.1) in terms of the covariances of X and E, denote, as before,

$$\Gamma_{xx} = E\{XX^T\} = \Gamma_{\mathrm{prior}},$$

$$\Gamma_{ee} = E\{EE^T\} = \Gamma_{\mathrm{noise}}.$$

and the cross covariances of X and E by

$$\Gamma_{xe} = \Gamma_{ex}^T = E\{XE^T\}. \tag{8.4}$$

Notice that the matrices (8.4) are often assumed to be zero as a consequence of the independence of X and E. In general, we obtain

$$\Gamma_{xb} = E\{X(AX+E)^T\}$$

$$= E\{XX^T\}A^T + E\{XE^T\} = \Gamma_{\mathrm{prior}} A^T + \Gamma_{xe}.$$

Similarly,

$$\Gamma_{bb} = \mathrm{E}\{(AX + E)(AX + E)^{\mathrm{T}}\}$$
$$= A\mathrm{E}\{XX^{\mathrm{T}}\}A^{\mathrm{T}} + A\mathrm{E}\{XE^{\mathrm{T}}\} + \mathrm{E}\{EX^{\mathrm{T}}\}A^{\mathrm{T}} + \mathrm{E}\{EE^{\mathrm{T}}\}$$
$$= A\Gamma_{\mathrm{prior}}A^{\mathrm{T}} + A\Gamma_{xe} + \Gamma_{ex}A^{\mathrm{T}} + \Gamma_{\mathrm{noise}}.$$

Therefore the expression for the midpoint and the covariance of the posterior density, i.e., the conditional mean and covariance,

$$x_{\mathrm{CM}} = \left(\Gamma_{\mathrm{prior}}A^{\mathrm{T}} + \Gamma_{xe}\right)\left(A\Gamma_{\mathrm{prior}}A^{\mathrm{T}} + A\Gamma_{xe} + \Gamma_{ex}A^{\mathrm{T}} + \Gamma_{\mathrm{noise}}\right)^{-1}b,$$

and

$$\Gamma_{\mathrm{post}} = \Gamma_{\mathrm{prior}}$$
$$-\left(\Gamma_{\mathrm{prior}}A^{\mathrm{T}} + \Gamma_{xe}\right)\left(A\Gamma_{\mathrm{prior}}A^{\mathrm{T}} + A\Gamma_{xe} + \Gamma_{ex}A^{\mathrm{T}} + \Gamma_{\mathrm{noise}}\right)^{-1}\left(A\Gamma_{\mathrm{prior}} + \Gamma_{ex}\right),$$

are more cumbersome than when using the formulas (8.2) and (8.3).

In the special case where X and E are independent and therefore the cross correlations vanish, the formulas simplify considerably,

$$x_{\mathrm{CM}} = \Gamma_{\mathrm{prior}}A^{\mathrm{T}}\left(A\Gamma_{\mathrm{prior}}A^{\mathrm{T}} + \Gamma_{\mathrm{noise}}\right)^{-1}b, \tag{8.5}$$

$$\Gamma_{\mathrm{post}} = \Gamma_{\mathrm{prior}} - \Gamma_{\mathrm{prior}}A^{\mathrm{T}}\left(A\Gamma_{\mathrm{prior}}A^{\mathrm{T}} + \Gamma_{\mathrm{noise}}\right)^{-1}A\Gamma_{\mathrm{prior}}. \tag{8.6}$$

The formulas (8.5) and (8.6) require that the prior and noise covariances are given. Sometimes - like in the case of smoothness priors - instead of these matrices, their *inverses* are available, and in some applications the inverse is sparse while the matrix itself is dense. Fortunately, there are also formulas in terms of the inverses. If X and E are independent and

$$B_{\mathrm{prior}} = \Gamma_{\mathrm{prior}}^{-1}, \quad B_{\mathrm{noise}} = \Gamma_{\mathrm{noise}}^{-1}.$$

we can write the posterior as

$$\pi(x \mid y) \propto \pi(x)\pi(y \mid x)$$
$$\propto \exp\left(-\frac{1}{2}x^{\mathrm{T}}B_{\mathrm{prior}}x - \frac{1}{2}(b - Ax)^{\mathrm{T}}B_{\mathrm{noise}}(b - Ax)\right).$$

Collecting the quadratic and linear terms above together, and completing the square we get

$$x^{\mathrm{T}}B_{\mathrm{prior}}x + (b - Ax)^{\mathrm{T}}B_{\mathrm{noise}}(b - Ax)$$
$$= x^{\mathrm{T}}\underbrace{\left(B_{\mathrm{prior}} + A^{\mathrm{T}}B_{\mathrm{noise}}A\right)}_{=B_{\mathrm{post}}}x - x^{\mathrm{T}}A^{\mathrm{T}}B_{\mathrm{noise}}b - b^{\mathrm{T}}B_{\mathrm{noise}}Ax + b^{\mathrm{T}}B_{\mathrm{noise}}b$$
$$= \left(x - B_{\mathrm{post}}^{-1}A^{\mathrm{T}}B_{\mathrm{noise}}b\right)^{\mathrm{T}}B_{\mathrm{post}}\left(x - B_{\mathrm{post}}^{-1}A^{\mathrm{T}}B_{\mathrm{noise}}b\right) + \underbrace{b^{\mathrm{T}}(\cdots)b}_{\text{independent of } x}.$$

Since the last term contributes only to the normalizing constant which is of no interest here, we write

$$\pi(x \mid y) \propto \exp\left(-\frac{1}{2}\left(x - B_{\text{post}}^{-1} A^{\text{T}} B_{\text{noise}} b\right)^{\text{T}} B_{\text{post}}\left(x - B_{\text{post}}^{-1} A^{\text{T}} B_{\text{noise}} b\right)\right),$$

from which we derive the alternative formulas

$$x_{\text{CM}} = B_{\text{post}}^{-1} A^{\text{T}} B_{\text{noise}} b = \left(B_{\text{prior}} + A^{\text{T}} B_{\text{noise}} A\right)^{-1} A^{\text{T}} B_{\text{noise}} b$$

$$= \left(\Gamma_{\text{prior}}^{-1} + A^{\text{T}} \Gamma_{\text{noise}}^{-1} A\right)^{-1} A^{\text{T}} \Gamma_{\text{noise}}^{-1} b, \tag{8.7}$$

and

$$\Gamma_{\text{post}} = B_{\text{post}}^{-1} = \left(\Gamma_{\text{prior}}^{-1} + A^{\text{T}} \Gamma_{\text{noise}}^{-1} A\right)^{-1}. \tag{8.8}$$

It is not immediately clear that these formulas are equivalent to (8.5) and (8.6). In fact, the proof requires a considerable amount of matrix manipulations and is left as an exercise.

EXAMPLE 8.1: Assume that both X and E are Gaussian white noise, i.e., $\Gamma_{\text{prior}} = \gamma^2 I$ and $\Gamma_{\text{noise}} = \sigma^2 I$. Then we have the two different expressions for the conditional mean:

$$x_{\text{CM}} = (A^{\text{T}} A + \delta^2 I)^{-1} A^{\text{T}} b = A^{\text{T}} (A A^{\text{T}} + \delta^2 I)^{-1} b, \quad \delta = \frac{\sigma}{\gamma}. \tag{8.9}$$

The former expression is already familiar: it is in fact the Tikhonov regularized solution of the linear system $Ax = b$. The latter is also a classical formula, the *Wiener filtered solution* of the linear system[1]. One may ask which formula should be used. If $A \in \mathbb{R}^{m \times n}$, since

$$A^{\text{T}} A + \delta^2 I \in \mathbb{R}^{n \times n}, \quad A A^{\text{T}} + \delta^2 I \in \mathbb{R}^{m \times m}.$$

at first it would seem that for underdetermined systems ($m < n$), the Tikhonov regularized solution is more appropriate while for overdetermined systems ($n < m$) one should favor Wiener filtering. The decision is not so clear cut, because the complexity of the computation depends on many other factors, than just the dimensionality of the problem, e.g., the sparsity and the structure of the matrices involved. Often it is advisable not to use *any* of the above formulas, but to notice that the posterior density is

$$\pi(x \mid b) \propto \exp\left(-\frac{1}{2\sigma^2}\|Ax - b\|^2 - \frac{1}{2\gamma^2}\|x\|^2\right)$$

$$= \exp\left(-\frac{1}{2\sigma^2}\left\|\begin{bmatrix} A \\ \delta I \end{bmatrix} x - \begin{bmatrix} b \\ 0 \end{bmatrix}\right\|\right).$$

---

[1] The formula goes back to the signal processing work of Norbert Wiener (1894–1964) from the fourties, and it is classically derived via projection techniques in probability space.

Therefore the conditional mean is the least squares solution of the system

$$\begin{bmatrix} A \\ \delta I \end{bmatrix} x = \begin{bmatrix} b \\ 0 \end{bmatrix},$$

which can often be computed effectively using iterative solvers.

Before applying the above results to our model problems, we discuss boundary effects that play an important role in inverse problems of imaging and signal analysis. In the following section, we show how the selection of suitable boundary conditions for a truncated signal or image can be dealt with from a statistical perspective, with techniques very similar to those introduced for the interpolation problem.

8.2 Aristotelian boundary conditions

To put boundary conditions into context, let us start by considering a one-dimensional deblurring problem, where the objective is to estimate the function $f : [0,1] \rightarrow \mathbb{R}$ from the observations of a noisy, blurred version of it, modelled as

$$g(s_j) = \int_0^1 \mathcal{A}(s_j, t) f(t) dt + e(s_j), \quad 1 \le j \le m,$$

with $\mathcal{A}$ a smoothing kernel. The prior belief is that the signal f is smooth. No particular information concerning the boundary behavior of f at the endpoints of the interval is available.

Discretize the model by dividing the support of the signal into n equal subintervals, let

$$x_j = f(t_j), \quad t_j = \frac{j}{n}, \quad 0 \le j \le n.$$

and consider the inverse problem of recovering x from the data.

To construct a prior density that favors non-oscillatory solutions, we proceed as in Section 7.2, and write a stochastic model,

$$X_j = \frac{1}{2}(X_{j-1} + X_{j+1}) + W_j, \quad 1 \le j \le n - 1,$$

where the innovation processes W_j are mutually independent, Gaussian random variables with zero mean and variance γ^2. In matrix form, the stochastic model can be written as

$$LX = W,$$

where $L \in \mathbb{R}^{(n-1) \times (n+1)}$ is the second order finite difference matrix (7.12), and

$$W \sim \mathcal{N}(0, \gamma^2 I), \quad W \in \mathbb{R}^{n-1}.$$

A natural candidate for the prior density for X, referred to as the *preprior* here, would be

$$\pi_{\mathrm{pre}}(x) \propto \exp\left(-\frac{1}{2\gamma^2}\|Lx\|^2\right) = \exp\left(-\frac{1}{2\gamma^2}x^{\mathrm{T}}Bx\right),$$

where

$$B = L^{\mathrm{T}}L \in \mathbb{R}^{(n+1)\times(n+1)}.$$

However, since it can be shown that[2]

$$\mathrm{rank}\,(B) = n - 1,$$

B is not invertible, hence π_{pre} is not a proper density in the sense that we cannot find a symmetric positive definite covariance matrix Γ such that

$$\exp\left(-\frac{1}{2\gamma^2}x^{\mathrm{T}}Bx\right) = \exp\left(-\frac{1}{2}x^{\mathrm{T}}\Gamma^{-1}x\right).$$

In the attempt to compensate for the rank-deficiency of B, let us see what happens if we augment L by adding a first and last row,

$$\frac{1}{2}\begin{bmatrix} 1 & -2 & 1 & & & \\ & 1 & -2 & 1 & & \\ & & \ddots & \ddots & \ddots & \\ & & & 1 & -2 & 1 \end{bmatrix} \rightarrow \frac{1}{2}\begin{bmatrix} -2 & 1 & & & & \\ 1 & -2 & 1 & & & \\ & 1 & -2 & 1 & & \\ & & \ddots & \ddots & \ddots & \\ & & & 1 & -2 & 1 \\ & & & & 1 & -2 \end{bmatrix} = L_{\mathrm{D}} \in \mathbb{R}^{(n+1)\times(n+1)}.$$

The augmented matrix L_{D} is symmetric positive definite, hence

$$\pi_{\mathrm{D}}(x) \propto \exp\left(-\frac{1}{2\gamma^2}\|L_{\mathrm{D}}x\|^2\right)$$

is a proper probability density.

To understand what the augmentation of L to L_{D} means from the point of view of statistical information, we look at random draws from this density. After making the change of variable

$$W = \frac{1}{\gamma}L_{\mathrm{D}}X,$$

and denoting by w a realization of W, we have that

$$\exp\left(-\frac{1}{2\gamma^2}\|L_{\mathrm{D}}x\|^2\right) = \exp\left(-\frac{1}{2}\|w\|^2\right),$$

that is, W is Gaussian white noise in $\mathbb{R}^{n+1}$. Therefore we can generate random draws from π_{D} by performing the following steps:

---

[2] This is not obvious, and the proof requires a fair amount of linear algebraic manipulations.

1. Draw a realization w of W,

$$w \sim \mathcal{N}(0, I), \quad (\mathtt{w} = \mathtt{randn(n+1,1)});$$

2. Find the corresponding realization x of X by solving the linear system

$$L_\mathrm{D} x = \gamma w$$

A few of these random draws are shown in Figure 8.1.

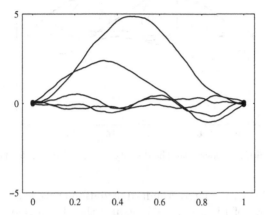

Fig. 8.1. Five random draws from the density π_D.

It is clear from Figure 8.1 that

- At the endpoints of the support all random draws almost vanish;
- The maximum variability in the draws seems to occur at the center of the support interval.

To understand why the endpoints of the random draws all approach zero, compute the *variance* of X at the jth pixel:

$$\mathrm{var}(X_j) = \mathrm{E}\{X_j^2\} = \int_{\mathbb{R}^{n+1}} x_j^2 \pi_\mathrm{D}(x) dx. \qquad (8.10)$$

Writing

$$x_j = e_j^\mathrm{T} x, \quad e_j = \begin{bmatrix} 0 & 0 & \cdots & 1 & \cdots & 0 \end{bmatrix}^\mathrm{T},$$

substituting it in (8.10) and remembering that, from our construction of the density, the inverse of the covariance matrix of X is $(1/\gamma^2)L_\mathrm{D}^\mathrm{T} L_\mathrm{D}$, we have that

$$\mathrm{var}(X_j) = e_j^\mathrm{T} \underbrace{\left(\int_{\mathbb{R}^n} xx^\mathrm{T} \pi_\mathrm{D}(x) dx \right)}_{=\mathrm{E}\{XX^\mathrm{T}\}} e_j$$

$$= \gamma^2 e_j^T \left(L_D^T L_D\right)^{-1} e_j = \gamma^2 e_j^T L_D^{-1} L_D^{-T} e_j$$
$$= \gamma^2 \|L_D^{-T} e_j\|^2.$$

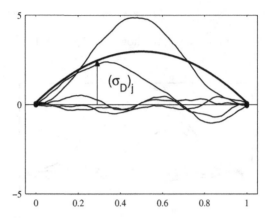

Fig. 8.2. Five random draws from the density π_D and the standard deviation curve.

In Figure 8.2 the calculated standard deviation curve is plotted along with the random draws of Figure 8.1. We deduce that by augmenting L into L_D we have implicitly forced the variance to nearly vanish at the endpoints. In fact, the construction of L_D is equivalent to extending x by zero outside the interval, modifying the stochastic model so that

$$X_0 = \frac{1}{2}X_1 + W_0 = \frac{1}{2}(\underbrace{X_{-1}}_{=0} + X_1) + W_0,$$

and

$$X_n = \frac{1}{2}X_{n-1} + W_n = \frac{1}{2}(X_{n-1} + \underbrace{X_{n+1}}_{=0}) + W_n.$$

While this may be very well justified in a number of applications and might have been supported by our prior belief about the solution, it is definitely not a universally good way to handle boundary points. In fact, if we do not know and we have no reason to believe that the signal approaches zero at the boundary, it is better to admit our ignorance[3] about the behavior of the solution at the boundary and let the data decide what is most appropriate.

To free the endpoints of the interval from assumptions that we cannot support, let's go back to $L \in \mathbb{R}^{(n-1)\times(n+1)}$. Instead of augmenting L into L_D, we use the idea of conditioning of Section 7.2 and proceed as follows:

---

[3] Quoting Stan Laurel's character in the classic movie *Sons of the Desert*, "Honesty [is] the best politics!"

- First express the believed smoothness at the interior points by the conditional density,

$$\pi(x_1, x_3, \ldots, x_{n-1} \mid x_0, x_n) \propto \exp\left(-\frac{1}{2\gamma^2}\|Lx\|^2\right),$$

conditioned on the values at the endpoints;
- Admit that we only *pretended* to know the values at endpoints, but *did not*;
- Free the endpoints by modelling them as random variables.

The conditional probability density of the interior points was derived in Section 7.2. After permuting and partitioning the components of x so that

$$x = \begin{bmatrix} x_1 \\ x_2 \\ \vdots \\ x_{n-1} \\ \hline x_n \\ x_0 \end{bmatrix} = \begin{bmatrix} x' \\ x'' \end{bmatrix},$$

and rearranging the matrix L accordingly,

$$L = \begin{bmatrix} L_1 & L_2 \end{bmatrix}, \quad L_1 \in \mathbb{R}^{(n-1)\times(n-1)}, \quad L_2 \in \mathbb{R}^{(n-1)\times2},$$

the conditional density can be written in the form

$$\pi(x' \mid x'') \propto \exp\left(-\frac{1}{2\gamma^2}\|L_1(x'' - L_1^{-1}L_2x'')\|^2\right).$$

Assume now that the boundary points X_0 and X_n are independent, zero mean Gaussian random variables with variance β^2, that is

$$\pi(x'') \propto \exp\left(-\frac{1}{2\beta^2}\|x''\|^2\right) = \exp\left(-\frac{1}{2}\|Kx''\|^2\right),$$

where

$$K = \frac{1}{\beta}I \in \mathbb{R}^{2\times2}.$$

As shown in Section 7.2, the density of X is

$$\pi(x) = \pi(x' \mid x'')\pi(x'') = \exp\left(-\frac{1}{2\gamma^2}\|L_A x\|^2\right),$$

where $L_A \in \mathbb{R}^{(n+1)\times(n+1)}$ is the invertible matrix

$$L_A = \begin{bmatrix} L_1 & L_2 \\ 0 & \delta K \end{bmatrix}, \quad \delta = \frac{\gamma}{\beta}.$$

Permuting the columns and rows of L_A to correspond the natural order of the pixels of x, $(1, 2, \ldots, n, 0) \to (0, 1, \ldots, n)$, we find that

$$
L_A = \frac{1}{2}
\begin{bmatrix}
2\delta & 0 & & & & \\
-1 & 2 & -1 & & & \\
 & -1 & 2 & -1 & & \\
 & & \ddots & \ddots & \ddots & \\
 & & & -1 & 2 & -1 \\
 & & & & 0 & 2\delta
\end{bmatrix}.
$$

If we want a prior density that expresses an approximately equal lack of knowledge[4] about the pixel values in the entire interval, the final step is to choose δ so that the variance is approximately the same at every point of the interval, i.e.,

$$\text{variance at } j = 0 \approx \text{variance at } j = [n/2] \approx \text{variance at } j = n,$$

where $[n/2]$ denotes the integer part of $n/2$. Observe that since the variance of X_j with respect to the prior density defined by the matrix L_A is

$$(\sigma_A)_j^2 = \gamma^2 \|L_A^{-T} e_j\|^2,$$

the condition above becomes

$$\|L_A^{-T} e_0\|^2 = \|L_A^{-T} e_{[n/2]}\|^2 = \|L_A^{-T} e_n\|^2.$$

To compute δ in an efficient manner, since away from the endpoints the variance of the random draws is fairly insensitive to our boundary assumptions, we can use the approximation

$$(\sigma_A)_j^2 \approx (\sigma_D)_j^2, \quad j = [n/2],$$

and set

$$\beta = (\sigma_D)_{n/2} = \gamma \|L_D^{-T} e_{[n/2]}\|,$$

to get

$$\delta = \frac{\gamma}{\beta} = \frac{1}{\|L_D^{-T} e_{[n/2]}\|}.$$

We call the prior obtained in this manner *Aristotelian boundary prior* because, like the Greek philosopher Aristotle[5] who believed that knowledge

---

[4] May be it would be more honest to say *ignorance* here also, but we opted for *lack of knowledge* to keep the moral high.

[5] Unfortunately Aristotle has a bad reputation in the scientific community, probably because many of his scientific speculations were eventually proved to be incorrect. Here we pay our tribute to Aristotle's philosophy of knowledge as a layered process, rather than to his scientific intuition. In fairness, though, it is safe to say that what we believe to be scientifically sound today will not hold any longer two thousand years from now, assuming that there is anybody to witness the consequences of today's sound science.

is a process which starts from a clean slate and is built up, layer after layer, as we experience nature, we also start with minimal assumptions about the boundary behavior of the solution and let the data determine what it should be.

Figure 8.3 shows a few random draws from the density obtained from the matrix L_A, together with the computed standard deviation curve. Note that now the variance of the values at the endpoints is similar to the variance of the values in the interior of the support. We remark that these random draws represent only what we believe a priori about our solution, before taking the measured data into consideration. Once we introduce the measured data via the likelihood, the values of the entries of the vector x will settle, in view of the prior belief *and* of the measurements.

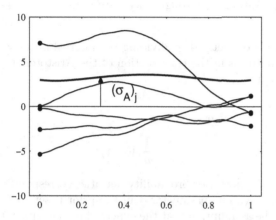

Fig. 8.3. Five random draws from the density π_A together with the standard deviation curve.

With the Aristotelian smoothness prior density in hand, we return to the deconvolution problem introduced in the beginning of this section. Let

$$y_j = g(s_j) = \int_0^1 \mathcal{A}(s_j, t) f(t) dt + e(s_j), \quad 1 \le j \le m,$$

where the blurring kernel is

$$\mathcal{A}(s, t) = c(s)t(1 - t)e^{-(t-s)^2/2w^2}, \quad w = 0.03,$$

and the scaling function c is chosen so that $\max_{t \in [0,1]}(K(s, t)) = 1$. In Figure 8.4 the kernel $t \mapsto \mathcal{A}(s, t)$ is plotted for two different values of s.

We assume that the sampling points are

$$s_j = (j - 1)/(m - 1), \quad m = 40.$$

and that the noise is normally distributed white noise with variance σ^2.

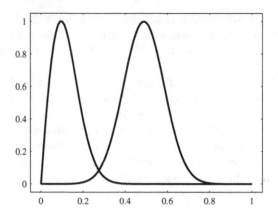

Fig. 8.4. The blurring kernels with $s = s_1$ and $s = s_{20}$.

The problem is discretized by dividing the interval into $n = 100$ subintervals of equal length, as in the construction of the Aristotelian prior, yielding a matrix equation

$$y = Ax + e,$$

where $A \in \mathbb{R}^{m \times (n+1)}$ has entries

$$A_{i,j} = \frac{1}{n} \mathcal{A}(s_i, t_j).$$

We compute the posterior probability densities corresponding to the prior densities π_{D} and π_{A}. The parameter γ in the prior is set to $\gamma = 1/n$, which corresponds to the assumption that the expected increments in the innovation process correspond to increments that are needed to cover a dynamic range of order of magnitude of unity. The noise level σ is assumed to be 3% of the noiseless signal.

In Figure 8.5 we have plotted the conditional means corresponding to both the Dirichlet (left panel) and the Aristotelian (right panel) smoothness priors, and the corresponding pointwise predictive output envelopes corresponding to two standard deviations. The true input is also plotted.

The results reveal several features of the priors. First, we observe that with the Dirichlet smoothness prior, the conditional mean and the credibility envelopes fail to follow the true input, which obviously is in conflict with the prior belief that it vanishes at the endpoints. The Aristotelian prior produces a better conditional mean estimate in this respect. On the other hand, the Aristotelian predictive envelope is considerably wider towards the end of the interval than in its interior. There is a natural explanation for this: by looking at the blurring kernels, we observe that they convey much less information about the signal at the endpoints of the interval than in the interior, therefore the posterior density at the endpoints is dominated by the prior. We

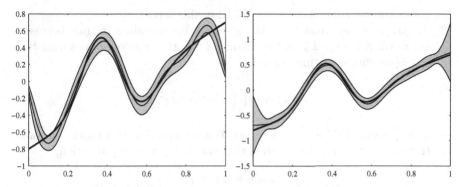

Fig. 8.5. The conditional means and two standard deviation predictive output envelopes using the Dirichlet smoothness prior (left) and the Aristotelian smoothness prior (right). The bold curve is the true input used to generate the data.

may conclude that the Dirichlet prior gives a false sense of confidence at the endpoints, while the Aristotelian prior admits the lack of information[6].

Exercises

This exercise shows how incorrect boundary values in the prior may have global effects on the solution of an inverse problem.

We consider the inverse Laplace transform: estimate the function f from the observations

$$g(s_j) = \int_0^\infty e^{-s_j t} f(t) dt + e_j, \quad 1 \le j \le m.$$

The Laplace transform is *notoriously* ill–conditioned.

1. Assume that the function f is supported on the interval $[0, 1]$, calculate the matrix A which is the discretized Laplace transform by using the piecewise constant approximation,

$$\int_0^1 e^{-s_j t} f(t) dt \approx \frac{1}{n} \sum_{k=1}^n e^{-s_j t_k} f(t_k), \quad t_k = \frac{k-1}{n-1}, \ 1 \le k \le n,$$

and choosing the data points s_j to be logarithmically equidistant, i.e.,

$$\log_{10}(s_j) = 4 \left(\frac{j-1}{m-1} - 1 \right), \quad 1 \le j \le m.$$

Calculate the condition number of the matrix A for different values of n.

---

[6] This example verifies the words of Thomas Gray: "Where ignorance is bliss, 'tis folly to be wise."

2. Calculate the L_A matrix as explained in this section.
3. Assume that you know that the signal f takes on values roughly between -2 and 2. Scale the L_A matrix in such a way that random draws from the corresponding smoothness prior,

$$\pi(x) \propto \exp\left(-\frac{1}{2}\alpha\|L_A x\|^2\right)$$

take on values in the target interval. Test your scaling with random draws.
4. Having set up the prior, assume that the true f is a step function,

$$f(t) = \begin{cases} 1, & \text{when } 0 \le t < a \\ 0, & \text{when } t \ge a \end{cases}, \quad 0 < a < 1.$$

Calculate analytically the noiseless data, then add Gaussian noise to it. Choose a low noise level, otherwise the results are rather bad.
5. Calculate the data also by using your matrix A and estimate the discrepancy that is due to the discretization. Is it larger or smaller than the additive arror? Notice that the level of the discretization error depends on the number of discretization points. The data that is generated by the forward model and is subsequently used to solve the inverse problem is often referred to as the *inverse crime data*, since it corresponds unrealistically well to the model one uses[7].
6. Calculate the conditional mean estimate and the covariance matrix. Plot the predictive output envelope of two posterior standard deviations.
7. Repeat the calculation using L_D instead of L_A. Notice that the wrong boundary value propagates also away from the boundary.
8. The step function corresponds badly to the smoothness prior assumption. Repeat the calculations with data that comes from a smooth function that takes on a non-zero value at the boundary.
9. Try the solutions with the inverse crime data. Do you notice any difference?
10. Show that the formulas (8.5)–(8.6) and (8.7)–(8.8) are equivalent. This may be a tedious task!

---

[7] In this book, we have not paid much attention to the issue of inverse crimes, albeit it is a very serious issue when model credibilities are assessed.

Sampling: the real thing

In Chapter 5, where we first introduced random sampling, we pointed out that sampling is used primarily to explore a given probability distribution and to calculate estimates for integrals via Monte Carlo integration. It was also indicated that sampling from a non-Gaussian probability density may be a challenging task. In this section we further develop the topic and introduce Markov Chain Monte Carlo (MCMC) sampling. We start the discussion with a small example concerning Markov chains.

EXAMPLE 9.1: Consider a small network consisting of *nodes* and *directed links* of the type that we could obtain as the result of an internet search: a set of keywords are given to a search engine that finds a set of internet sites containing those keywords. The sites found are the nodes of the network and each site might contain links to other sites. If the node A contains a link to the site B which is also a node in the network, the nodes are linked to each others by a directed link from A to B. Observe that the site B need not have a link to site A. Figure 9.1 shows a small prototype of such a network.

If we want to list the network nodes in a descending order of importance, as some search engines do, then the natural question which arises is whether it is possible to deduce from the network structure alone which node is the most important.

At first one might think that the most important node is the one most referenced by the other nodes. But the question is more complex: in fact, there may be several insignificant nodes referring to one particular link, while there are several prominent ones referring to each other, forming thus a large cluster of relatively well referenced nodes. Therefore a better definition of importance could be based on *random walk*: when surfing the network randomly, always going from one node to another by selecting one of the directed links randomly, the most visited node would seem to be the most important one. Notice that the visits are done randomly, since we were not allowed to pay attention to the contents of the sites.

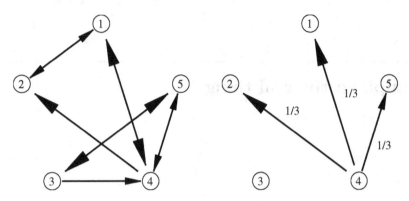

Fig. 9.1. The network (left) and the transition probabilities for moving from node 4 (right).

To explore the network, we write a *transition matrix*, defining the probabilities of transition between the nodes. In the right panel of Figure 9.1 we consider the probabilities of different moves from node 4. Since there are three links starting from this node, the transition probabilities of each one of them is equal to 1/3. The transition matrix corresponding to this network is

$$
P = \begin{bmatrix}
0 & 1 & 0 & 1/3 & 0 \\
1/2 & 0 & 0 & 1/3 & 0 \\
0 & 0 & 0 & 0 & 1/2 \\
1/2 & 0 & 1/2 & 0 & 1/2 \\
0 & 0 & 1/2 & 1/3 & 0
\end{bmatrix}.
$$

The kth column of P contains the transition probabilities from the kth node. Hence, if we start from node 1, the probability density of the next state is

$$
\pi_2 = P\pi_1 = \begin{bmatrix} 0 \\ 1/2 \\ 0 \\ 1/2 \\ 0 \end{bmatrix}, \quad \pi_1 = \begin{bmatrix} 1 \\ 0 \\ 0 \\ 0 \\ 0 \end{bmatrix}.
$$

Observe that the choice of π_1 reflects the fact that initially we are at node 1 with certainty. Similarly, the probability density after n steps is

$$
\pi_n = P\pi_{n-1} = P^2\pi_{n-2} = \cdots = P^n\pi_1.
$$

Define a sequence of random variables $X_n \in \mathbb{R}^5$ as follows: the components of X_n all vanish except one, and the index of the non-vanishing component is drawn from the density π_n. Hence, in the previous example, the possible realizations of the random variable X_2 are

$$x_2 = \begin{bmatrix} 0 \\ 1 \\ 0 \\ 0 \\ 0 \end{bmatrix} = e_2 \quad \text{or} \quad x_2 = \begin{bmatrix} 0 \\ 0 \\ 0 \\ 1 \\ 0 \end{bmatrix} = e_4,$$

and

$$P\{X_2 = e_2\} = P\{X_2 = e_4\} = 1/2.$$

If we want to move one step further, we have

$$\pi_3 = P^2 \pi_1 = \begin{bmatrix} 2/3 \\ 1/6 \\ 0 \\ 0 \\ 1/6 \end{bmatrix}.$$

These are the probabilities of the third state, assuming that we know that the sequence starts at node 1, i.e.,

$$\pi(x_3 \mid x_1) = \pi_3, \quad x_1 = e_1.$$

But what if we know also the second state? Suppose that $X_2 = x_2 = e_2$. Then, evidently, the probability distribution of X_3 becomes

$$\pi(x_3 \mid x_1, x_2) = Pe_2 = e_1, \quad x_1 = e_1, \ x_2 = e_2.$$

The relevant point here is that this result does not depend on the value of X_1: whichever way we came to the second node $X_2 = e_2$, the probability density of X_3 is always Pe_2, that is,

$$\pi(x_3 \mid x_1, x_2) = \pi(x_3 \mid x_2).$$

More generally, if we know the state $X_n = x_n$, the next state has density Px_n regardless of the more remote past states. We may write

$$\pi(x_{n+1} \mid x_1, x_2, \ldots, x_n) = \pi(x_{n+1} \mid x_n). \tag{9.1}$$

We say that a *discrete time stochastic process* $\{X_1, X_2, \ldots\}$ is a *Markov process* if it has the property (9.1). This condition is often expressed by saying that "tomorrow depends on the past only through today". We have encountered previously Markov processes when constructing smoothness priors. There the Markov property referred to spatial neighbors instead of temporal predecessors.

Consider now the question of how to assess the importance of the nodes. One way is to start a realization of the Markov chain at some node, say, $x_1 = e_1$, generate a sample,

$$S = \{x_1, x_2, \ldots, x_N\},$$

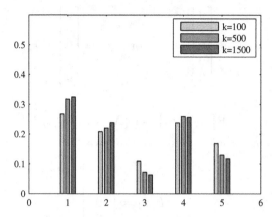

Fig. 9.2. The relative visiting frequencies at different nodes with three different chain lengths N.

by drawing x_j from the probability density Px_{j-1} and compute the visiting frequency of the different nodes.

Figure 9.2 shows, in histogram form, the percentages of visitations of the different nodes,

$$h_j = \frac{1}{N} \#\{x_k \mid x_k = e_j\},$$

for different values of N. The histograms seem to indicate that the first node is the most important, according to this criterion.

Finally, let us search for a more efficient way of reaching the result that we obtained above by random walk. Assume that, as the number of iterations grows, the probability distribution converges towards a limit distribution,

$$\lim_{n \to \infty} P^n \pi_1 = \pi_\infty.$$

Is there a simple way to find the limit distribution? By writing

$$\pi_{n+1} = P\pi_n,$$

and taking the limit on both sides, we obtain

$$\pi_\infty = P\pi_\infty. \tag{9.2}$$

In other words, the limit distribution, if it exists, is the normalized eigenvector of the transition matrix P corresponding to the unit eigenvalue. We say that the limit distribution is an *invariant distribution* of the transition matrix P, since it remains unchanged under the action of P. Thus, to classify the node according to their importance, it suffices to find the eigenvector corresponding to the eigenvalue one. In Figure 9.3, we have plotted the components of this normalized eigenvector and the visiting histogram corresponding to $N = 1500$ random steps. The matching between the values of the components of the eigenvector and the relative frequencies of visitation of the different nodes is quite striking, and it increases with N.

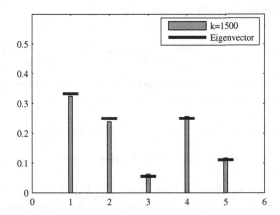

Fig. 9.3. The visiting histogram with $N = 1500$ and the components of the eigenvector corresponding to eigenvalue one.

It is worth mentioning that since one is the largest eigenvalue of the transition matrix, its corresponding eigenvector can be computed by the power method. The most popular web search engines order the results of their search in a similar fashion.

In the previous example, we started with the transition matrix P and wanted to find the associated invariant probability distribution. Suppose now that π_∞ is known instead, and that we want to generate a sample distributed according to π_∞. A natural way to proceed would be to generate a realization of the Markov chain using the matrix P starting from some initial point. After a few iterations, the elements of the chain will be distributed according to π_∞. The plan is perfect except for a small detail: we have access to π_∞ by assumption but *we don't know P!* Taking care of this detail is indeed the core question in Markov Chain Monte Carlo!

Since, in general, we are interested in problems in which the random variables take on values in $\mathbb{R}^n$, we are now going to set the stage for Markov Chain Monte Carlo with non-discrete state space. We start by defining the concept of *random walk* in $\mathbb{R}^n$ which, as the name suggests, is a process of moving around by taking random steps. The most elementary random walk can be defined as follows:

1. Start at a point of your choice $x_0 \in \mathbb{R}^n$.
2. Draw a random vector $w_1 \sim \mathcal{N}(0, I)$ and set $x_1 = x_0 + \sigma w_1$.
3. Repeat the process: Set $x_{k+1} = x_k + \sigma w_{k+1}$, $w_{k+1} \sim \mathcal{N}(0, I)$.

Using the random variables notation, the location of the random walk at time k is a realization of the random variable X_k, and we have an evolution model

$$X_{k+1} = X_k + \sigma W_{k+1}, \quad W_{k+1} \sim \mathcal{N}(0, I).$$

The conditional density of X_{k+1}, given $X_k = x_k$, is

$$\pi(x_{k+1} \mid x_k) = \frac{1}{(2\pi\sigma^2)^{n/2}} \exp\left(-\frac{1}{2\sigma^2}\|x_k - x_{k+1}\|^2\right) = q(x_k, x_{k+1}).$$

The function q is called the *transition kernel* and it is the continuous equivalent of the transition matrix P in the discrete state space example. Since

$$q(x_0, x_1) = q(x_1, x_2) = \cdots = q(x_k, x_{k+1}) = \cdots,$$

i.e., the step is always equally distributed independently of the value of k, the kernel is called *time invariant*.

Clearly, the process above defines a *chain* $\{X_k, \ k = 0, 1, \cdots\}$ of random variables, each of them having a probability density of its own. The chain is a discrete time stochastic process with values in $\mathbb{R}^n$ and has the particular feature that the probability distribution of each variable X_{k+1} depend on the past only through the previous element X_k of the chain. This can be expressed in terms of the conditional densities as

$$\pi(x_{k+1} \mid x_0, x_1, \ldots, x_k) = \pi(x_{k+1} \mid x_k).$$

As in the discrete case of the previous example, a stochastic process with this property is called a *Markov chain*.

EXAMPLE 9.2: To understand the role of the transition kernel, consider a Markov chain defined by a random walk model in $\mathbb{R}^2$,

$$X_{k+1} = X_k + \sigma W_{k+1}, \quad W_{k+1} \sim \mathcal{N}(0, C), \tag{9.3}$$

where $C \in \mathbb{R}^{2\times 2}$ is a symmetric positive definite matrix, whose eigenvalue decomposition we write as

$$C = UDU^{\mathrm{T}}. \tag{9.4}$$

The inverse of C can be decomposed as

$$C^{-1} = UD^{-1}U^{\mathrm{T}} = \underbrace{\left(UD^{-1/2}\right)\left(D^{-1/2}U^{\mathrm{T}}\right)}_{=L},$$

so the transition kernel can be written as

$$q(x_k \mid x_{k+1}) = \pi(x_{k+1} \mid x_k) \propto \exp\left(-\frac{1}{2\sigma^2}\|L(x_k - x_{k+1})\|^2\right).$$

We may write the random walk model (9.3) in an alternative way as

$$X_{k+1} = X_k + \sigma L^{-1} W_{k+1}, \quad W_{k+1} \sim \mathcal{N}(0, I), \tag{9.5}$$

where the random step is whitened.
To demonstrate the effect of the covariance matrix, let

$$U = \begin{bmatrix} \cos\theta & -\sin\theta \\ \sin\theta & \cos\theta \end{bmatrix}, \quad \theta = \frac{\pi}{3},$$

and

$$D = \mathrm{diag}(s_1^2, s_2^2), \quad s_1 = 1, \ s_2 = 5.$$

In the light of the random walk model (9.5), the random steps have a component about five times larger in the direction of the second eigenvector e_2 than in the first eigenvector e_1, where

$$e_1 = \begin{bmatrix} \cos\theta \\ \sin\theta \end{bmatrix}, \quad e_2 = \begin{bmatrix} -\sin\theta \\ \cos\theta \end{bmatrix}.$$

The left panel of Figure 9.4 shows three random walk realizations with the covariance matrix $C = I$ starting from the origin of $\mathbb{R}^2$ and choosing the step size $\sigma = 0.1$. In the right panel, the covariance matrix is chosen as above. Obviously, by judiciously choosing the transition kernel, we may guide the random walk quite effectively.

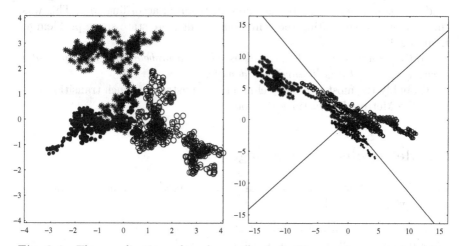

Fig. 9.4. Three realizations of random walks. Left: The covariance matrix of the random step is $\sigma^2 I$, with the standard deviation $\sigma = 0.1$, and the number of steps in each realization is $N = 500$. Right: The covariance is C given in (9.4). The line segments mark the eigendirections of the covariance matrix.

Consider now an arbitrary transition kernel q. Assume that X is a random variable whose probability density is given. To emphasize the particular role of this density, we denote it by $\pi(x) = p(x)$. Suppose that we generate a new random variable Y by using a given kernel $q(x, y)$, that is,

$$\pi(y \mid x) = q(x, y).$$

The probability density of this new variable Y is found by marginalization,

$$\pi(y) = \int \pi(y \mid x)\pi(x)dx = \int q(x, y)p(x)dx.$$

If the probability density of Y is equal to the probability density of X, i.e.,

$$\int q(x, y)p(x)dx = p(y),$$

we say that p is an *invariant density* of the transition kernel q. The classical problem in the theory of Markov chains can then be stated as follows: *Given a transition kernel, find the corresponding invariant density.*

When using Markov chains to sample from a given density, we are actually considering the *inverse problem*: Given a probability density $p = p(x)$, generate a sample that is distributed according to it. If we had a transition kernel q with invariant density p, generating such sample would be easy: starting from x_0, draw x_1 from $q(x_0, x_1)$, and repeat the process starting from x_1. In general, given x_k, draw x_{k+1} from $q(x_k, x_{k+1})$. After a while, the x_k's generated in this manner are more and more distributed according to p. This was the strategy for generating the sample in the discrete state space problem of Example 9.1.

So the problem we are facing now is: *Given a probability density p, find a kernel q such that p is its invariant density.*

Probably the most popular technique for constructing such transition kernel is the Metropolis–Hastings method.

9.1 Metropolis–Hastings algorithm

Consider the following, more general, Markov process: starting from the current point $x \in \mathbb{R}^n$, either

1. Stay put at x with probability $r(x)$, $0 \le r(x) < 1$, or
2. Move away from x using a transition kernel $R(x, y)$.

Since R by definition is a transition kernel, the mapping $y \mapsto R(x, y)$ defines a probability density, hence

$$\int_{\mathbb{R}^n} R(x, y)dy = 1.$$

Denoting by $\mathcal{A}$ the event of moving away from x, and by $\neg\mathcal{A}$ the event of not moving, the probabilities of the two events are

$$P\{\mathcal{A}\} = 1 - r(x), \quad P\{\neg\mathcal{A}\} = r(x).$$

Now, given the current state $X = x$, we want to know what is the probability density of Y generated by the above strategy. Let $B \subset \mathbb{R}^n$, and consider the probability of the event $Y \in B$. It follows that

$$P\{Y \in B \mid X = x\} = P\{Y \in B \mid X = x, \mathcal{A}\}P\{\mathcal{A}\}$$
$$+ P\{Y \in B \mid X = x, \neg\mathcal{A}\}P\{\neg\mathcal{A}\},$$

in other words, Y can end up in B either through a move or by staying put. The probability of arriving in B through a move is obviously

$$P\{Y \in B \mid X = x, \mathcal{A}\} = \int_B R(x, y)dy.$$

On the other hand, arriving in B without moving can only happen if x already is in B. Therefore,

$$P\{Y \in B \mid X = x, \neg\mathcal{A}\} = \chi_B(x) = \begin{cases} 1, \text{ if } x \in B, \\ 0, \text{ if } x \notin B \end{cases},$$

where χ_B is the characteristic function of B.

In summary, the probability of reaching B from x is

$$P\{Y \in B \mid X = x\} = (1 - r(x)) \int_B R(x, y)dy + r(x)\chi_B(x).$$

Since we are only interested in the probability of $Y \in B$, we marginalize over x and calculate the probability of $Y \in B$ *regardless of the initial position*,

$$P\{Y \in B\} = \int P\{Y \in B \mid X = x\}p(x)dx$$

$$= \int p(x) \left(\int_B (1 - r(x))R(x, y)dy \right) dx + \int \chi_B(x)r(x)p(x)dx$$

$$= \int_B \left(\int p(x)(1 - r(x))R(x, y)dx \right) dy + \int_B r(x)p(x)dx$$

$$= \int_B \left(\int p(x)(1 - r(x))R(x, y)dx + r(y)p(y) \right) dy.$$

Recalling that, by definition,

$$P\{Y \in B\} = \int_B \pi(y)dy,$$

the probability density of Y must be

$$\pi(y) = \int p(x)(1 - r(x))R(x,y)dx + r(y)p(y).$$

Our goal is then to find a kernel R such that $\pi(y) = p(y)$, that is

$$p(y) = \int p(x)(1 - r(x))R(x,y)dx + r(y)p(y),$$

or, equivalently,

$$(1 - r(y))p(y) = \int p(x)(1 - r(x))R(x,y)dx. \tag{9.6}$$

For simplicity, let us denote

$$K(x,y) = (1 - r(x))R(x,y),$$

and observe that, since R is a transitional kernel,

$$\int K(y,x)dx = (1 - r(y))\int R(y,x)dx = 1 - r(y).$$

Therefore, condition (9.6) can be expressed in the form

$$\int p(y)K(y,x)dx = \int p(x)K(x,y)dx,$$

which is called the *balance equation*. This condition is satisfied, in particular, if the integrands are equal,

$$p(y)K(y,x) = p(x)K(x,y).$$

This condition is known as the *detailed balance equation*. The Metropolis-Hastings algorithm is simply a technique of finding a kernel K that satisfies it.

The Metropolis-Hastings algorithm starts by selecting a *proposal distribution*, or *candidate generating kernel* $q(x,y)$, chosen so that generating a Markov chain with it is easy. It is mainly for this reason that a Gaussian kernel is a very popular choice.

If q satisfies the detailed balance equation, i.e., if

$$p(y)q(y,x) = p(x)q(x,y),$$

we let $r(x) = 0$, $R(x,y) = K(x,y) = q(x,y)$, and we are done, since the previous analysis shows that p is an invariant density for this kernel. If, as is more likely to happen, the detailed balance equation is not satisfied, the left hand side is larger or smaller than the right hand side. Assume, for the sake of definiteness, that

$$p(y)q(y,x) < p(x)q(x,y). \tag{9.7}$$

To enforce the detailed balance equation we can define the kernel K to be

$$K(x,y) = \alpha(x,y)q(x,y),$$

where α is a suitable correcting factor, chosen so that

$$p(y)\alpha(y,x)q(y,x) = p(x)\alpha(x,y)q(x,y). \tag{9.8}$$

Since the kernel α need not be symmetric, we can choose

$$\alpha(y,x) = 1,$$

and let the other correcting factor be determined from (9.8):

$$\alpha(x,y) = \frac{p(y)q(y,x)}{p(x)q(x,y)} < 1.$$

Observe that if the sense of inequality (9.7) changes, we simply interchange the roles of x and y, letting instead $\alpha(x,y) = 1$. In summary, we define K as

$$K(x,y) = \alpha(x,y)q(x,y), \quad \alpha(x,y) = \min\left\{1, \frac{p(y)q(y,x)}{p(x)q(x,y)}\right\}.$$

The expression for K looks rather complicated and it would seem that generating random draws should be all but simple. Fortunately, the drawing can be performed in two phases, as in the case of rejection sampling, according to the following procedure:

1. Given x, draw y using the transition kernel $q(x,y)$.
2. Calculate the *acceptance ratio*,

$$\alpha(x,y) = \frac{p(y)q(y,x)}{p(x)q(x,y)}.$$

3. Flip the α–coin: draw $t \sim \text{Uniform}([0,1])$; if $\alpha > t$, accept y, otherwise stay put at x.

To show why this works, let's consider, again, the probability that the proposal is accepted,

$$\pi(\mathcal{A} \mid y) = \text{probability of acceptance of } y, \text{ which is } = \alpha(x,y),$$

and

$$\pi(y) = \text{probability density of } y, \text{ which is } = q(x,y).$$

Then, from the Bayes formula,

$$\begin{aligned} \pi(y,\mathcal{A}) &= \text{probability density of accepted } y\text{'s} \\ &= \pi(\mathcal{A} \mid y)\pi(y) = \alpha(x,y)q(x,y) \\ &= K(x,y), \end{aligned}$$

which is exactly what we wanted to prove.

We will now identify some convenient features of the algorithm while following it into action.

EXAMPLE 9.3: Consider the following probability density in $\mathbb{R}^2$,

$$\pi(x) \propto \exp\left(-\frac{1}{2\sigma^2}((x_1^2 + x_2^2)^{1/2} - 1)^2 - \frac{1}{2\delta^2}(x_2 - 1)^2\right), \qquad (9.9)$$

where

$$\sigma = 0.1, \quad \delta = 1,$$

whose equiprobability curves are shown in Figure 9.5.

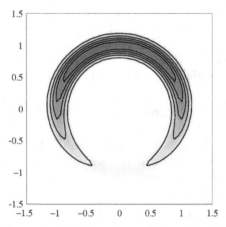

Fig. 9.5. The equiprobability curves of the original density (9.9)

We first explore this density with a random walk sampler. For this purpose consider the white noise random walk proposal,

$$q(x, y) = \frac{1}{\sqrt{2\pi\gamma^2}}\exp\left(-\frac{1}{2\gamma^2}\|x - y\|^2\right).$$

When the transition kernel is symmetric, i.e.,

$$q(x, y) = q(y, x),$$

the Metropolis-Hastings algorithm is particularly simple, because

$$\alpha(x, y) = \frac{\pi(y)}{\pi(x)}.$$

In this case the acceptance rule can be stated as follows: if the new point has higher probability, accept it immediately; if not, accept with probability determined by the ratio of the probabilities of the new point and the old one.

We start the Markov chain from the origin, $x_0 = (0,0)$, and to illustrate how it progresses, we adopt the plotting convention that each new accepted point is plotted by a dot. If a proposal is rejected and we remain in the current position, the current position is plotted with a larger dot, so that the area of the dot is proportional to the number of rejections. A Matlab code that generates the sample can be written as follows:

```
nsample = 500;              % Size of the sample
Sample = zeros(2,nsample);
count = 1;
x = [0;0];                  % Initial point
Sample(:,1) = x;
lprop_old = logpdf(x);  % logpdf = log of the
prob.density

while count < nsample
        % draw candidate
        y = x + step*randn(2,1);
        % check for acceptance
        lprop =  logpdf(y);
        if lprop - lprop_old > log(rand);
            % accept
            accrate = accrate + 1;
            x = y;
            lprop_old = lprop;
        end
        count = count+1;
        Sample(:,count+1) = x;
    end
```

We remark that in the actual computation, the acceptance ratio is calculated in logarithmic form: we accept the move $x \to y$ if

$$\log p(y) - \log p(x) > \log t, \quad t \sim \text{Uniform}([0,1]).$$

The reason for proceeding in this manner is to avoid numerical problems with overflow or underflow in the computation of the ratio of $p(y)$ and $p(x)$.

In our first exploration of the density, we decide to move rather conservatively, taking very small steps by setting $\gamma = 0.02$. A plot of the first 500 points generated by the algorithm is shown in Figure 9.6.

It is clear from the plot that, after 500 draws, the sampler has not even started to explore the density. In fact, almost the entire sample is needed to move from the initial point to the numerical support of the density. This initial tail, which has nothing to do with the actual probability density, is

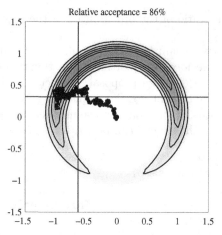

Fig. 9.6. The Metropolis-Hastings sample with step size $\gamma = 0.02$. The sample mean is marked by the cross hair.

usually referred to as the *burn-in* of the sample. It is normal procedure in MCMC sampling methods to discard the beginning of the sample to avoid that the burn-in affects the estimates that are subsequently calculated from the sample. In general, it is not easy to decide a priori how many points should be discarded.

The second observation is that the *acceptance rate* is high: almost nine out of ten proposed moves are accepted. A high acceptance rate usually indicates that the chain is moving too conservatively, and longer steps should be used to get better coverage of the distribution.

Motivated by the observed high acceptance rate, we then increase the step by a factor of hundred, choosing $\gamma = 2$. The results of this modification can be seen in Figure 9.7. Now the acceptance rate is only 7%, meaning that most of the time the chain does not move and the proposals are rejected. Notice that the big dots in the figure indicate points from which the chain does not want to move away. The burn-in effect is practically absent, and the estimated conditional mean is much closer to the actual mean what one could expect. The result seems to suggest that too low acceptance rate is better than too high.

By playing with the steplength, we may be able to *tune* the proposal distribution so as to achieve an acceptance rate between 20% and 30%, which is often advocated as optimal.

Figure 9.8 shows the results obtained with $\gamma = 0.5$, yielding an acceptance rate of approximately 26%. We see that the points are rather well distributed over the support of the probability density. The estimated mean, however, is not centered, indicating that the size of the sample is too small.

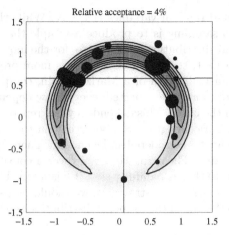

Fig. 9.7. The Metropolis-Hastings sample with the step size $\gamma = 2$. The computed sample mean is marked by the cross hair.

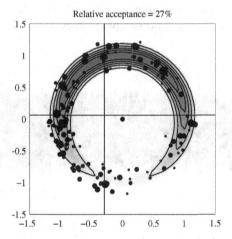

Fig. 9.8. The Metropolis-Hastings sample with the step size $\gamma = 0.5$.

The previous example shows that the choice of the proposal distribution has an effect on the quality of the sample thus generated. While in the two dimensional case it is fairly easy to assess the quality of the sampling strategy by simply looking at the scatter plot of the sample, in higher dimensions this approach becomes impossible, and more systematic means are needed to analyze the sample. No definitive measures for the quality can be given, and we merely graze this rather complex topic here.

The Central Limit Theorem gives an idea of where to look for a measure of the quality of the sample. Remember that according to the Central Limit Theorem, the asymptotic convergence rate of a sum of N independently sampled, identically distributed random variables is $1/\sqrt{N}$. While the goal of Markov Chain Monte Carlo sampling is to produce a sample that is asymptotically drawn from the limit distribution, taking care, for the most part, of the *identically distributed* aspect, the part *independent* is more problematic. Clearly, since the sample is a realization of a Markov chain, complete independency of the sample points cannot be expected: every draw depends at least on the previous element in the chain. This dependency has repercussions on the convergence of Monte Carlo integrals. Suppose that on the average, only every kth sample point can be considered independent. Then, by the asymptotic law of the Central Limit Theorem, we may expect a convergence rate of the order of $\sqrt{k/N}$, a rate that is painfully slow if k is large. Therefore, when designing the Metropolis-Hastings strategies, we should be particularly careful, in choosing the step length in the proposal distribution so that the *correlation length* is small.

Let us begin with visual inspection of samples. Suppose that we have a sampling problem in n spatial dimensions. While we cannot inspect a scatter plot, we may always look at the *sample histories* of individual component, plotting each individual component as a function of the sample index. The first question is then what a good sample history looks like.

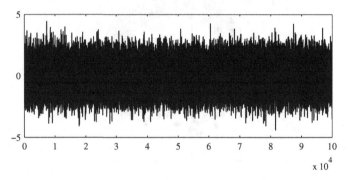

Fig. 9.9. A Gaussian white noise signal of length 100 000.

A typical example of a sample with completely uncorrelated elements is a *white noise signal*: at each discrete time instant, the sample is drawn independently. A Gaussian white noise signal is, in fact, a realization of a Gaussian multivariate vector with covariance $\sigma^2 I$. Each component, independent of the remaining ones, represents a realization of one time instance. Figure 9.9 shows a realization of a white noise signal. It looks like a "fuzzy worm". This gives a visual description for a MCMCer of what is meant by saying that good sample histories should look like fuzzy worms. This visual description, although rather vague, is in fact, quite useful to quickly assess the quality of a sample.

To obtain a more quantitative measure of the quality of a sample, we may look at its correlation structure. The *autocorrelation* of a signal is a useful tool to analyze the independency of realizations. Let z_j, $0 \leq j \leq N$ denote a discrete time finite segment of a signal. Assume, for simplicity, that the signal has zero mean. After augmenting the signal with trailing zeroes to an infinite signal, $0 \leq j < \infty$, consider the discrete convolution,

$$h_k = \sum_{j=0}^{\infty} z_{j+k} z_j, \quad k = 0, 1, 2 \ldots$$

When $k = 0$, the above formula returns the total energy of the signal,

$$h_0 = \sum_{j=1}^{N} z_j^2 = \|z\|^2.$$

If z is a white noise signal and $k > 0$, the random positive and negative contributions cancel out, and $h_k \approx 0$. This observation gives a natural tool to analyze the independency of the components in a sample: plot

$$k \mapsto \widehat{h}_k = \frac{1}{\|z\|^2} h_k, \quad k = 0, 1, \ldots$$

and estimate the correlation length from the rate of decay of this sequence. The quantity $\widehat{h}_k$ is the autocorrelation of z with *lag k*.

In the following example we demonstrate how to use of this idea.

EXAMPLE 9.4: Consider the horseshoe distribution of the previous example, and the Metropolis–Hastings algorithm with a white noise proposal distribution. We consider three different step sizes, $\gamma = 0.02$ which is way too small, $\gamma = 0.8$ which is in a reasonable range and, finally, $\gamma = 5$, which is clearly too large. We generate a sample $\{x_1, x_2, \ldots, x_N\}$ of size $N = 100\,000$, calculate the mean,

$$\overline{x} = \frac{1}{N} \sum_{j=1}^{N} x_j,$$

and the lagged autocorrelations of the centered components,

$$\widehat{h}_{i,k} = \frac{1}{\|z\|^2} \sum_{j=1}^{N-k} z_{j+k} z_j, \quad z_j = (x_j - \overline{x})_i, \quad i = 1, 2.$$

In Figure 9.10 we have plotted the sample history of the first component with the different step sizes. Visually, the most reasonable step size produces a sample history which is similar to the white noise sample, while with the smallest step size there is a strong low frequency component that indicates a slow walk around the density. The too large step size, on the other hand, causes the sampler to stay put for long periods of times with an adverse effect on the independency of the samples.

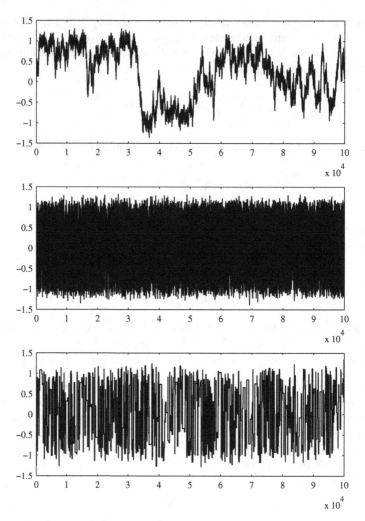

Fig. 9.10. The sample histories of the first component of our random variable with different step sizes in the proposal distribution. From top to bottom, $\gamma = 0.02$, $\gamma = 0.8$ and $\gamma = 5$.

In Figure 9.11, the corresponding autocorrelations for both components are shown. This figure confirms the fact that the correlation length, or the period after which the sample points can be considered insignificantly correlated, is much shorter with the step length $\gamma = 0.8$ than in the extreme cases.

In the previous examples, we used Gaussian white noise random walks as proposal distributions. It is quite clear that this proposal takes very poorly into account the fact that the density that we are sampling from is very

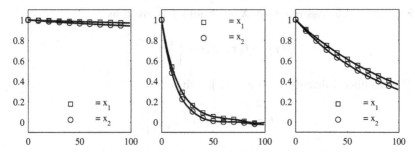

Fig. 9.11. The autocorrelations of the components with the different proposal step sizes. From left to right, $\gamma = 0.02$, $\gamma = 0.8$ and $\gamma = 5$.

anisotropic. In the next example we show how the shape of the density can be taken into account when designing the proposal distribution.

Consider the problem of estimating the random variable X from the observation

$$B = f(X) + E, \quad E \sim \mathcal{N}(0, \sigma^2 I),$$

where f is a non-linear function and we assume a priori that X is a Gaussian and independent of E. For simplicity, let $X \sim \mathcal{N}(0, \gamma^2 I)$. The posterior density in this case is

$$\pi(x \mid b) \sim \exp\left(-\frac{1}{2\gamma^2}\|x\|^2 - \frac{1}{2\sigma^2}\|b - f(x)\|^2\right).$$

and we want to explore it by using the Metropolis-Hastings algorithm.

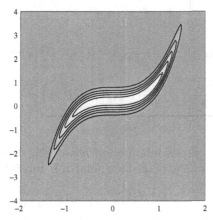

Fig. 9.12. The equiprobability curves of likelihood density $x \mapsto \pi(b_{\text{observed}} \mid x)$ in Example 9.5.

EXAMPLE 9.5: Assume that $X \in \mathbb{R}^2$ and f is given by

$$f(x) = f(x_1, x_2) = \begin{bmatrix} x_1^3 - x_2 \\ x_2/5 \end{bmatrix}.$$

The likelihood density, $x \mapsto \pi(y \mid x)$, with

$$b = b_{\text{observed}} = \begin{bmatrix} -0.2 \\ 0.1 \end{bmatrix},$$

has been plotted in Figure 9.12. If the variance γ^2 of the prior is large, the likelihood and the posterior density are essentially equal.

Since the standard Gaussian white noise proposal density does not take into account the fact that the numerical support of the posterior density is shaped like a relatively narrow ridge, it proposes steps with equally long components, on average, along the ridge and perpendicularly to it. Although for the present two-dimensional problem, a white noise random walk proposal might still be performing well, in more realistic, high-dimensional problems, this proposal could be unacceptably expensive, since long proposal steps would be almost always rejected, while short steps would explore the distribution too slowly. An alternative way would be to use *local* Gaussian approximation of the density as proposal. Assuming that x is the current point, begin with writing a linear approximation for f,

$$f(x + w) \approx f(x) + Df(x)w,$$

from which we construct a quadratic approximation of the exponent of the posterior density,

$$\frac{1}{2\gamma^2} \|x + w\|^2 + \frac{1}{2\sigma^2} \|b - f(x + w)\|^2$$

$$\approx \frac{1}{2\gamma^2} \|x + w\|^2 + \frac{1}{2\sigma^2} \|b - f(x) - Df(x)w\|^2$$

$$= w^{\mathrm{T}} \underbrace{\left(\frac{1}{2\gamma^2} I + \frac{1}{2\sigma^2} (Df(x))^{\mathrm{T}} Df(x) \right)}_{=H(x)} w + \text{lower order terms}.$$

The matrix $H(x)$ carries information about the shape of the posterior density in a neighborhood of the base point x. Figure 9.13 shows the ellipses

$$w^{\mathrm{T}} H(x) w = \text{constant},$$

calculated at two different base points.
We can now use

$$q(x, y) \propto \exp\left(-\frac{1}{2\delta^2} (x - y)^{\mathrm{T}} H(x)(x - y) \right)$$

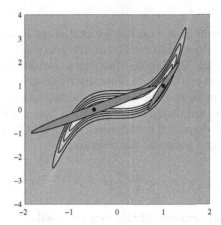

Fig. 9.13. Variance ellipsoids of the proposal distribution at two different points. The step control parameter in this plot is $\delta = 0.2$, i.e., we have $w^{\mathrm{T}} H(x) w = \delta^2 = 0.04$ at the boundary of each ellipse.

as a proposal distribution, but we need to remember that the distribution is no longer symmetric, since the matrix H depends on x. Therefore, in the process of updating x we compute the factorization

$$H(x) = L(x)^{\mathrm{T}} L(x),$$

and we write

$$y = x + \delta L(x)^{-1} w, \quad w \sim \mathcal{N}(0, I).$$

The tuning parameter $\delta > 0$ can be used to adjust the step size so as to achieve the desired acceptance rate. Furthermore, to calculate the acceptance ratio α, we need to evaluate the differential $Df(y)$ at the proposed point y, since

$$q(y, x) \propto \exp\left(-\frac{1}{2\delta^2} (y - x)^{\mathrm{T}} H(y)(y - x)\right).$$

We remark that if y is accepted, the value of the differential can be used for the subsequent linear approximation, based at y.

The process of updating the proposal distribution while the sampling algorithm moves on is called *adaptation*. The adaptation strategy explained in the previous example is suitable for low dimensional problems, but usually too costly for high dimensional problems and, in practice, the adaptation is normally based on the sample that has been generated so far: the empirical covariance of the sample is calculated and used as the covariance of the proposal distribution, see the exercises at the end of this chapter. A deeper discussion of different variants of MCMC algorithms is beyond the scope of this book.

Exercises

1. Write a Matlab code based on the adaptation of Example 9.5. Compare the acceptance, autocorrelation and convergence speed of the algorithm with a random walk Metropolis-Hastings algorithm where the proposal distribution is a Gaussian white noise distribution.

2. To get an idea of more realistic adaptive methods, try the following: Using a white noise proposal distribution, generate first a sample $\{x_1, \ldots, x_N\}$, then calculate the empirical covariance matrix C of this sample. Continue the sampling using a proposal distribution

$$q(x - y) \propto \exp\left(-\frac{1}{2}(x - y)^{\mathrm{T}}C^{-1}(x - y)\right).$$

Investigate the performance of this new proposal.

Wrapping up: hypermodels, dynamic priorconditioners and Bayesian learning

In the previous chapters we have seen how priors give us a way to bring into numerical algorithms our belief about the solution that we want to compute. Since the most popular distributions used as priors depend themselves on parameters, a natural question to ask is how these parameters are chosen. While in some cases the parameter choice might be natural and easy to justify, in some of the applications where the subjective framework gives big improvements over a deterministic approach, the values which should be chosen for these parameters is not clear, and therefore we regard them as random variables. The procedure of introducing a distribution for the parameters defining the prior is sometimes referred to as hypermodel in the statistics literature. Depending on the complexity of the problem, it may happen that the parameters defining the distribution of the parameters of the priors are also modelled as random variables, thus introducing a nested chain of hypermodels.

To illustrate this, let's consider first an example from signal processing.

EXAMPLE 10.1: Assume that we want to recover a signal that is known to be continuous, except for a possible jump discontinuity at a known location. The natural question which arises is how we can design a prior which conveys this belief.

More specifically, let $f : [0, 1] \to \mathbb{R}$ denote the continuous signal, and let

$$x_j = f(t_j), \quad t_j = \frac{j}{n}, \quad 0 \le j \le n,$$

denote its discretization. For simplicity, let us assume that $f(0) = x_0 = 0$. We denote the vector of the unknown components by x,

$$x = \begin{bmatrix} x_1 \\ \vdots \\ x_n \end{bmatrix} \in \mathbb{R}^n.$$

A good candidate for a prior that favors signals with small variations is a Gaussian first order smoothness prior,

$$\pi_{\text{prior}}(x) \propto \exp\left(-\frac{1}{2\gamma^2}\|Lx\|^2\right),$$

where the matrix L is the first order finite difference matrix,

$$L = \begin{bmatrix} 1 & & & \\ -1 & 1 & & \\ & \ddots & \ddots & \\ & & -1 & 1 \end{bmatrix} \in \mathbb{R}^{n\times n}.$$

This prior corresponds to a Markov model,

$$X_j = X_{j-1} + W_j, \quad W_j \sim \mathcal{N}(0, \gamma^2), \tag{10.1}$$

that is, each discrete value of the signal is equal to the previous value, up to a white innovation process that has a fixed variance γ^2. The value of the variance γ^2 of the innovation term is a reflection of how large a jump we expect the signal could have.

If we believe that in the interval $[t_{k-1}, t_k]$ the signal could have a much larger jump than at other locations, it is therefore natural to replace the Markov model (10.1) at $j = k$ with a model with a larger variance,

$$X_k = X_{k-1} + W_k, \quad W_k \sim \mathcal{N}\left(0, \frac{\gamma^2}{\delta^2}\right),$$

where $\delta < 1$ is a parameter controlling the variance of the kth innovation. It is easy to verify that this modification leads to a prior model

$$\pi_{\text{prior}}(x) \propto \exp\left(-\frac{1}{2\gamma^2}\|D^{1/2}Lx\|^2\right),$$

where $D^{1/2} \in \mathbb{R}^{n\times n}$ is a diagonal matrix,

$$D^{1/2} = \text{diag}(1, 1, \cdots, \delta, \cdots, 1), \quad \delta < 1,$$

the location of δ being fixed and known.

To see the effect of this modification of the standard smoothness prior model, we generate random draws from this distribution, setting $\gamma = 1$ and $n = 100$, meaning that over each subinterval except the kth one, we expect random increments of the order ~ 1. We then modify the kth standard deviation, $k = 40$, to be equal to $1/\delta$. In Figure 10.1, we show four random draws from the prior with values $\delta = 0.1$ (left) and $\delta = 0.02$ (right). The outcomes confirm that the expected size of the jump is controlled by the size of the parameter $1/\delta$, since

$$E\{(X_{k+1} - X_k)^2\} = \left(\frac{\gamma}{\delta}\right)^2.$$

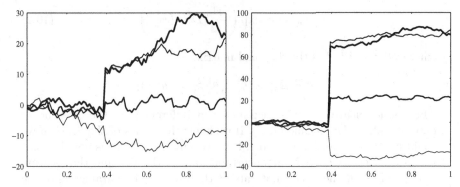

Fig. 10.1. Random draws from the prior distribution with $\delta = 0.1$ (left) and $\delta = 0.02$ (right) as the control parameter of the step size in the subinterval around $t = 0.4$.

The construction of the prior in the previous example was simple, because the prior information was given in a *quantitative* way: We believed *a priori* that

1. the signal could have exactly one jump,
2. the location of the possible jump was known within the resolution of the discretized support of the signal, and
3. an estimate for the range of the size of the jump, that is, its variance, was available.

A much more challenging, and undoubtedly more common situation in practical applications, is that the only prior information available is inherently qualitative. For instance, a geophysicist doing electromagnetic profiling of the Earth may summarize what is known a priori about the electric conductivity profile as a function of depth with a statement of the type: "I'm expecting to see a mildly varying conductivity profile, but some discontinuities may occur." Thus the prior belief is clearly qualitative, since it is not expressed in terms of numbers and mathematical formulas. The counterpart in our example would be to expect a smooth signal, which could have a few jumps of unknown magnitude at some unknown locations. The challenge is now to build a prior that allows the *data* to decide how many, how large and where the jumps are. To see how we can proceed, we revisit Example 10.1.

EXAMPLE 10.2: In Example 10.1 we assumed that the signal increments $X_j - X_{j-1}$ were equally distributed except for one. We can allow all increments to be different by replacing (10.1) with

$$X_j = X_{j-1} + W_j, \quad W_j \sim \mathcal{N}\left(0, \frac{1}{\theta_j}\right), \quad \theta_j > 0.$$

This leads to a prior of the form

$$\pi_{\text{prior}}(x) \propto \exp\left(-\frac{1}{2}\|D^{1/2}Lx\|^2\right),\tag{10.2}$$

where $D^{1/2} \in \mathbb{R}^{n \times n}$ is the diagonal matrix,

$$D^{1/2} = \text{diag}\left(\theta_1^{1/2}, \theta_2^{1/2}, \cdots, \theta_n^{1/2}\right).$$

This model clearly accepts increments of different size, the sizes being determined by the parameters θ_j. If we knew the values of the θ_j, then we could proceed as in Example 10.1. However, in general this is not the case, thus we have just traded our uncertainty about the size and the location of the jumps for the uncertainty about the value of the parameters of the prior.

Among the questions raised by this example, a particularly intriguing - and interesting - one is what happens if we treat the parameters θ_j of the prior as variables, and whether they could be determined by the data. Before answering, we need to be concerned about a quantity that normally does not interest us, namely the normalizing constant of the prior density. We start with computing the integral of the exponential on the right side of equation (10.2). Letting

$$I = \int_{\mathbb{R}^n} \exp\left(-\frac{1}{2}\|D^{1/2}Lx\|^2\right) dx$$

$$= \int_{\mathbb{R}^n} \exp\left(-\frac{1}{2}x^T L^T DLx\right) dx,$$

and

$$R = D^{1/2}L,$$

we have

$$I = \int_{\mathbb{R}^n} \exp\left(-\frac{1}{2}\|Rx\|^2\right) dx.$$

Since the matrix $R \in \mathbb{R}^{n \times n}$ is invertible, the change of variable

$$Rx = z, \quad |\det(R)|dx = dz$$

leads to the formula

$$I = \frac{1}{|\det(R)|} \int_{\mathbb{R}^n} \exp\left(-\frac{1}{2}\|z\|^2\right) dz$$

$$= \frac{(\sqrt{2\pi})^n}{|\det(R)|}.$$

The determinant of R can be expressed in terms of the parameters θ_j, since

$$\det(L^T DL) = \det(R^T R) = \det(R)^2,$$

and therefore the normalized prior density is

$$\pi_{\text{prior}}(x) = \sqrt{\frac{\det(L^{\mathsf{T}} D L)}{(2\pi)^n}} \exp\left(-\frac{1}{2}\|D^{1/2} L x\|^2\right).$$

The dependency of the normalizing constant on the parameter vector θ is implicitly given via the determinant. It is possible to solve the explicit dependency, but since the computation is rather complicated, we skip it here. The point that we want to make here is that the price that we pay to treat the parameter θ also as an unknown along with x is that the normalizing constant, that depends on θ, cannot be neglected.

Let's now express the fact that the prior depends on the parameter θ by writing it as the conditional density

$$\pi_{\text{prior}}(x \mid \theta) \propto \det(L^{\mathsf{T}} D L)^{1/2} \exp\left(-\frac{1}{2}\|D^{1/2} L x\|^2\right), \quad D^{1/2} = \operatorname{diag}(\theta)^{1/2},$$
(10.3)

where the notation emphasizes the idea that the right side is the prior *provided* that the prior parameter θ is known.

Before further pursuing the idea of determining both x and θ from the data, we discuss the computation of the determinant in the normalizing constant.

Consider the random variables W_j that define the innovations, or random increments, in the Markov model for the signal. They are assumed to be mutually independent, hence their probability density, assuming the variances known, is

$$\pi(w) \propto \prod_{j=1}^{n} \exp\left(-\frac{1}{2}\theta_j w_j^2\right) = \exp\left(-\frac{1}{2}\|D^{1/2} w\|^2\right).$$

Unlike the normalizing constant of the density of x, the normalizing constant of this density has a simple explicit expression in terms of the entries of the parameter vector θ. Indeed, since D is diagonal and the determinant of a diagonal matrix is simply the product of its diagonal elements, we have that

$$\pi(w \mid \theta) \propto (\theta_1 \theta_2 \cdots \theta_n)^{1/2} \exp\left(-\frac{1}{2}\|D^{1/2} w\|^2\right)$$

$$= \exp\left(-\frac{1}{2}\|D^{1/2} w\|^2 + \frac{1}{2}\sum_{j=1}^{n} \log \theta_j\right).$$

Furthermore, since it is easy to construct a bijection between the increments w_j and the discretized signal values x_j by letting

$$w_j = x_j - x_{j-1}, \quad x_0 = 0,$$

or, in matrix notation,

$$w = Lx,$$

where L is the first order finite difference matrix. Conversely, the signal can be recovered from the increments by observing that

$$x_j = \sum_{k=1}^{j} w_k,$$

or, in the matrix notation,

$$x = Bw,$$

where

$$B = L^{-1} = \begin{bmatrix} 1 & & & \\ 1 & 1 & & \\ \vdots & & \ddots & \\ 1 & 1 & \ldots & 1 \end{bmatrix}.$$

The immediate consequence of this observation is that, from the point of view of information, it does not make any difference whether we express the prior belief in terms of the signal or in terms of its increments. However, writing the signal in terms of the increments simplifies considerably the expression for the prior. For this reason, we make the increments the unknowns of primary interest.

The next issue that needs to be addressed is what is believed of the prior for the parameters θ_j that will be treated themselves as random variables. Consider a Gaussian smoothness prior for the increments and write a Bayesian *hypermodel*,

$$\pi(w,\theta) = \pi_{\text{prior}}(w \mid \theta)\pi_{\text{hyper}}(\theta),$$

where the *hyperprior* $\pi_{\text{hyper}}(\theta)$ expresses our belief about θ. In the case of our signal example, a reasonable hyperprior would allow some of the θ_j to deviate strongly from the average, a situation which would correspond to the qualitative information that the signal might contain jumps. This suggests a fat tailed probability density that allows outliers, such as the exponential density with a positivity constraint

$$\pi_{\text{hyper}}(\theta) \propto \pi_+(\theta)\exp\left(-\frac{\gamma}{2}\sum_{j=1}^{n}\theta_j\right),$$

where $\pi_+(\theta)$ is one if all components of θ are positive, and vanishes otherwise, and $\gamma > 0$ is a hyperparameter.

Assuming that we have a linear observation model of the form

$$g(s_j) = \mathcal{A}f(t_j) + e_j, \qquad 1 \le j \le m,$$

whose discretized version is

$$b = Ax + e = ABw + e, \qquad A \in \mathbb{R}^{m \times n},$$

where the additive noise e is white Gaussian with variance σ^2, the likelihood model becomes

$$\pi(b \mid w) \propto \exp\left(-\frac{1}{2\sigma^2}\|b - ABw\|^2\right).$$

Thus, from the Bayes formula it follows that the posterior distribution is of the form

$$\pi(w, \theta \mid b) \propto \pi(b \mid w)\pi(w, \theta)$$

$$\propto \exp\left(-\frac{1}{2\sigma^2}\|b - ABw\|^2 - \frac{1}{2}\sum_{j=1}^{n}\theta_j w_j^2 - \frac{\gamma}{2}\sum_{j=1}^{n}\theta_j + \frac{1}{2}\sum_{j=1}^{n}\log\theta_j\right).$$

10.1 MAP estimation or marginalization?

In general, once we have an expression for the posterior, we can proceed to estimate the variable of primary interest by computing either the Maximum A Posteriori estimate or the Conditional Mean estimate.

In the literature, a third way, first estimating the prior parameters from marginal distributions and subsequently using them to estimate the unknowns of primary interest, is often advocated. More specifically, the third way in this case would amount to

1. Calculating the marginal distribution

$$\pi(\theta \mid y) = \int \pi(w, \theta \mid y)dw,$$

   often referred to as the *evidence* of the parameter θ;
2. Estimating θ by maximizing the evidence $\pi(\theta \mid y)$,

$$\widehat{\theta} = \mathrm{argmax}\pi(\theta \mid y);$$

3. Estimating w from the conditional density

$$\pi(w \mid \widehat{\theta}, y) \propto \pi(w, \widehat{\theta} \mid y).$$

Technical reasons make it cumbersome to implement the third way[1]. In fact, since both the prior and the likelihood of w are Gaussian, after some algebraic manipulations we arrive at the following analytic expression for the marginal density of θ,

---

[1] Like in the political scene of the last century, the expression third way indicates a break from the two leading ways. Historical third ways include the Frankfurt School and eurocommunism. Their implementation was also cumbersome for technical reasons.

$\pi(\theta \mid b)$

$$\propto \int \exp\left(-\frac{1}{2\sigma^2}\|b - ABw\|^2 - \frac{1}{2}\sum_{j=1}^n \theta_j w_j^2 - \frac{\gamma}{2}\sum_{j=1}^n \theta_j + \frac{1}{2}\sum_{j=1}^n \log\theta_j \right) dw$$

$$\propto \frac{1}{\det(\sigma^2(AB)^{\mathsf{T}}AB + D)^{1/2}} \exp\left(-\frac{\gamma}{2}\sum_{j=1}^n \theta_j + \frac{1}{2}\sum_{j=1}^n \log\theta_j \right),$$

where $D = \mathrm{diag}(\theta)$. Unfortunately this expression requires the evaluation of a determinant, a step that we want to avoid. Therefore we resort to the first way, namely to the computation of the Maximum A Posteriori estimate.

To compute the Maximum A Posteriori estimate, we first write the posterior in the form

$$\pi(w, \theta \mid b) \propto \exp\left(-\frac{1}{2}\|b - ABw\|^2 - \frac{1}{2}\|D^{1/2}w\|^2 - \frac{\gamma}{2}\sum_{j=1}^n \theta_j + \frac{1}{2}\sum_{j=1}^n \log\theta_j \right),$$

then observe that the Maximum A Posteriori estimate is the minimizer of the functional

$$F(w, \theta) = \frac{1}{2}\left\| \begin{bmatrix} AB \\ D^{1/2} \end{bmatrix} w - \begin{bmatrix} b \\ 0 \end{bmatrix} \right\|^2 - \frac{1}{2}\sum_{j=1}^n \log\theta_j + \frac{\gamma}{2}\sum_{j=1}^n \theta_j,$$

which can be found by solving alternatingly two simple minimization problems:

1. Fix the current estimate of the parameter of the prior, $\theta = \theta_c$, and update w by minimizing the first term of $F(w, \theta_c)$. The new value of w is the solution, in the least squares sense, of the linear system

$$\begin{bmatrix} AB \\ D^{1/2} \end{bmatrix} w = \begin{bmatrix} b \\ 0 \end{bmatrix}, \quad D = \mathrm{diag}(\theta_c).$$

2. Given the current estimate of $w = w_c$, update θ by minimizing $F(w_c, \theta)$. The new value of θ is the solution of the equation

$$\nabla_\lambda F(w_c, \theta) = 0,$$

for which we have the analytic expression

$$\theta_j = \frac{1}{w_{c,j}^2 + \gamma}.$$

Since $1/\gamma$ is an upper bound for θ_j, the hyperparameter γ effectively bounds from above the prior parameters θ_j. In the following example we apply this algorithm to a deblurring problem, and discuss the influence of the hyperparameter γ on the results.

EXAMPLE 10.3: Consider a one-dimensional deconvolution problem. Let I be the interval $I = [0, 3]$, and $a(s, t)$ a Gaussian blurring kernel

$$a(s, t) = \exp\left(-\frac{1}{2\delta^2}(s - t)^2\right), \qquad s, t \in I,$$

whose width is controlled by the parameter δ. We discretize the integral by dividing the interval of integration into $n = 150$ identical subintervals and set the width of the blurring so that the full width half value (FWHV) is 50 pixels by choosing

$$\delta = \frac{50}{2\sqrt{2 \log 2n}}.$$

and add Gaussian white noise with standard deviation 5% of the maximum of the noiseless blurred signal. We test the algorithm on a piecewise constant signal, shown along with the blurred noisy data in Figure 10.2.

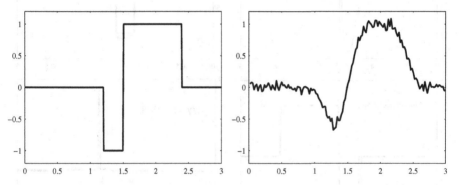

Fig. 10.2. The true input signal (left) and the blurred noisy signal (right). The noise level is 5% of the maximum of the noiseless signal.

Figure 10.3 shows the results obtained with the algorithm described above, as well as the approximate solutions computed during the iterations. The initial values for the θ_j were chosen as

$$\theta_{0,j} = \frac{1}{\gamma}, \qquad \gamma = 10^{-5}.$$

It is clear that after a few iterations the entries of θ away from the jumps take on very high values, and that the method has no problems with detecting the discontinuities even from a strongly blurred and rather noisy

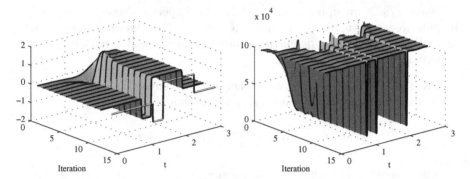

Fig. 10.3. The evolution of the sequential iterations of the signal (left) and the prior parameter θ (right). The value of the hyperparameter is $\gamma = 10^{-5}$.

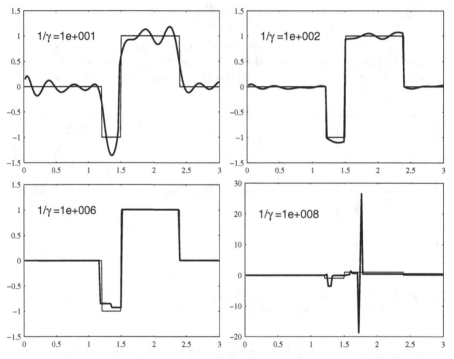

Fig. 10.4. The MAP estimates computed with various values of γ. The two estimates in the top row correspond to values of γ which are too large, while in the estimate on the bottom right the parameter is too small. Observe the difference in the scale of the bottom right figure.

signal. Figure 10.4 illustrates the role the value of the hyperparameter γ: although the range of suitable values of γ is rather wide, ranging from 10^{-7} to 10^{-3}, a large γ restricts the dynamical range of the parameters θ_j excessively and the solution remains too smooth, while when γ is too small the probability density of the smoothness parameters θ_j behaves as an improper density and the MAP estimation fails. The effect of γ on the quality of the reconstructed signal is more critical when the amount of noise is large, hence the information about the location of the jumps is harder to extract from the data.

The previous example conforms to the rule of thumb that, while the choice of the hyperparameter is important, the method is rather robust since the interval of reasonable values covers several orders of magnitudes.

Observe that one could have arrived at the above iterative method via a different reasoning: after each update w, we re-evaluate the prior information and update the prior density by changing θ. This process is sometimes referred to as *Bayesian learning*.

10.2 Bayesian hypermodels and priorconditioners

The computation of the Maximum A Posteriori estimate using a Bayesian hypermodel can be recast in terms of priorconditioners. This reformulation is particularly attractive for large scale problems, where the updating of w using direct methods may become unfeasible, or if the matrix A is not given explicitly. In the general setting, right priorconditioners have been introduced in the approximation of the Maximum A Posteriori estimate using iterative methods for the solution of the associated linear system in the least squares sense. Since the right priorconditioner comes from the factorization of the inverse of the covariance matrix of the prior, in the case of hypermodels it will have to be updated each time a new estimate of the parameters of the prior becomes available. This implies that, in the computation of the Maximum A Posteriori estimate, we solve repeatedly a linear system of equations by iterative methods using a right preconditioner which conveys the latest information about the underlying prior.

More specifically, let D_c be the diagonal matrix constructed from the current approximation θ_c of θ and consider the problem of minimizing

$$\frac{1}{2}\|ABw - b\|^2 + \frac{1}{2}\|D_c^{1/2}w\|^2$$

with respect to w. Since $D_c^{1/2}$ is invertible, we can introduce

$$z = D_c^{1/2}w, \qquad w = D_c^{-1/2}z,$$

and compute an approximate solution z_+ of the linear system

$$ABD_{\mathrm{c}}^{-1/2}z = b$$

by an iterative linear system solver. Letting

$$w_+ = D_{\mathrm{c}}^{-1/2}z_+$$

and

$$x_+ = Bw_+,$$

we can update the parameters of the prior according to the formula

$$\theta_{+,j} = \frac{1}{w_{+,j}^2 + \gamma},$$

and compute a new priorconditioner $D_+^{1/2} = \mathrm{diag}\{\theta_{+,1}, \ldots, \theta_{+,n})$, to be used in the iterative solution of the linear system. It should be pointed out that the ill-conditioning of the linear system that we solve and the presence of noise in the right hand side require that we stop iterating before the amplified noise components dominate the computed solution, and that we use an iterative solver suited for numerically inconsistent systems. We finally remark that the updating of the priorconditioner in the context of Bayesian hyperpriors is philosophically in line with the idea of flexible preconditioning for the GMRES method, where the change of the preconditioner reflects new acquired knowledge about the system. We conclude by revisiting Example 10.1, where we now use priorconditioned iterative methods for the restoration of the blurred and noisy signal.

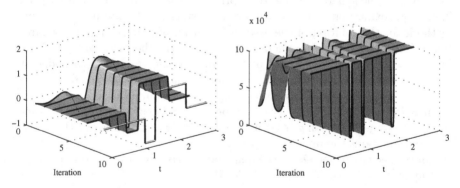

Fig. 10.5. The evolution of the dynamically priorconditioned CGLS iterations of the signal (left) and the prior parameter θ (right).

Given the blurred and noisy signal of Example 10.1, we use the CGLS iterative method for the solution of the linear system

$$ABw = b$$

with right preconditioner

$$D^{1/2} = \text{diag}\{\theta_1^{1/2}, \ldots, \theta_n^{1/2}\}.$$

In our computation we use the CGLS algorithm, allowing up to 15 iteration steps to reduce the residual below a tolerance of 10^{-3}. At the end of each CGLS sweep we use the new approximation of the signal increments to update the θ_j, hence the associated priorconditioner. The actual number of iteration steps needed to attained the desired residual reduction remains consistently below the maximum allowed, and decreases with each sweep. Figure 10.5 shows that the evolution of the computed solution is analogous to that obtained when computing the Maximum A Posteriori estimate directly.

References

[GR96] Gilks, W.R., Richardson, S. and Spiegelhalter, D.J.: Markov Chain Monte Carlo in Practice. Chapman&Hall/CRC Press, Boca Raton (1996)

[Gh96] Ghahramani, S.: Fundamentals of Probability. Prentice Hall (1996)

[GV89] Golub, G. and van Loan, C.F.:Matrix Computations. The John Hopkins University Press, Baltimore (Third edition) (1989).

[Ha98] Hansen, P.C.: Rank-Deficient and Discrete Ill-Posed Problems. SIAM, Philadelphia (1998)

[Je04] Jeffrey, R.: Subjective Probability: The Real Thing. Cambridge University Press, Cambridge (2004)

[Jo02] Joliffe, I.T.: Principal Component Analysis. Springer Verlag, New York (Second edition) (2002)

[KS05] Kaipio, J. and Somersalo, E.: Statistical and Computational Inverse Problems. Springer Verlag, New York (2005)

[PT01] Press, S.J. and Tanur, J.M.: The Subjectivity of Scientists and the Bayesian Approach. John Wiley&Sons, Inc., New York (2001)

[Sa03] Saad, Y.: Iterative Methods for Sparse Linear Systems. SIAM, Philadelphia (Second edition) (2003)

[Ta05] Tarantola, A.: Inverse Problem Theory. SIAM, Philadelphia (New edition) (2005)

[TB97] Trefethen, L.N. and Bau, D.: Numerical Linear algebra. SIAM, Philadelphia (1997)

Index

A–conjugate direction, 71
$A^{\mathrm{T}}A$–conjugate direction, 72
α–coin, 104, 171

absolutely continuous, 7
acceptance rate, 103, 174
acceptance ratio, 104, 171
adaptation, 181
aliasing, 89
anomalies, 118, 122
approximate solution, 70
Arnoldi process, 74, 75
Arnoldi relation, 76
autocorrelation, 176
autoregressive Markov model, 52
averaging, 22

bacteria, 4, 51, 94
balance equation, 170
bandwidth, 44
Bayes formula, 10, 55
Bayesian probability, 3
burn-in, 174

calibration, 29
candidate generating kernel, 170
CCD, Charged Coupled Device, 14, 17
Central Limit Theorem, 16, 29, 92, 121, 175
CG algorithm, 72
CGLS algorithm, 73
Cholesky decomposition, 93
Cholesky factorization, 57, 63
condition number, 78, 81

conditional covariance, 134
conditional density, 56
conditional expectation, 11
conditional mean, 11, 134
Conditional Mean estimate, 57, 189
conditional probability, 6
conditional probability density, 10, 48
conditioning, 40
confidence, 25
Conjugate Gradient method, 70
Conjugate Gradient method for Least Squares, 72
convergence criterion, 70
convergence rate, 108
convolution, 67
correlation coefficient, 9
correlation length, 176
counting process, 14
covariance, 9
covariance matrix, 11
credibility, 25
credibility ellipse, 25
credibility interval, 38
cumulative distribution function, 99

deconvolution, 81, 114, 123, 157, 191
detailed balance equation, 170
direct menthod, 63
discrepancy, 73
Discrete Fourier Transform (DFT), 68
discrete time stochastic process, 163

eigenvalue decomposition, 23, 36, 78, 110, 124

empirical Bayes methods, 51
empirical covariance, 23
empirical mean, 23
equiprobability curves, 24, 25, 27
error function, 101
error function, inverse of, 101
event, 6
evidence, 189
expectation, 8
expectation, Poisson process, 15
extrapolation, 144

false positive, 58, 125
Fast Fourier Transform (FFT), 67
finite difference matrix, first order, 115,
    184, 188
finite difference matrix, second order,
    54, 135
formally determined problem, 61
frequentist statistics, 3
fudge factor, 81

Gaussian approximation, 17, 19, 24, 45,
    48, 119, 180
Gaussian distribution, 16, 32
Gaussian elimination, 63
General Minimal Residual method, 74
GMRES algorithm, 76
Golden Rule, 100
gravitational force, 144

hard subspace constraint, 121
Hessenberg matrix, 76
hidden parameter, 35
hyperparameter, 188, 192

ignorance, 22, 154
ill-conditioned matrix, 81
ill-conditioning, 77
ill-determined rank, 119
independent events, 6
independent random variables, 9
independent realizations, 31
independent, identically distributed
    (i.i.d.), 16, 29, 47, 52
innovation process, 52
interpolation, 134
invariant density, 168
invariant distribution, 164

inverse crime, 160
Inverse Cumulative Distribution Rule,
    100
Inverse Discrete Fourier Transform
    (IDFT), 89
inverse problem, 1, 41, 49, 55
inverse problem, linear, 147
iterative method, 63
iterative solver, 67

Jacobian, 46
joint probability density, 9, 40

knowledge, 48
Krylov subspace, 70
Krylov subspace iterative methods, 70,
    108
Kullback–Leibler distance, 18

lack of information, 3, 39
Laplace transform, 159
Law of Large Numbers, 22, 23, 30, 33,
    34, 92
least squares solution, 62, 64, 76, 78
likelihood, 41
likelihood function, 32
limit distribution, 164
linear system, 56, 58, 61, 67, 107, 111,
    132, 153, 193, 194
log-likelihood function, 32
log-normal distribution, 38, 47

marginal density, 10, 55
marginalization, 40
Markov chain, 166
Markov Chain Monte Carlo, 99, 161
Markov process, 163
Matlab code, 27, 53, 69, 81, 87, 96, 97,
    101, 102, 141, 173
Maximum A Posteriori estimate, 115,
    189
Maximum A Posteriori estimator, 57,
    84
Maximum Likelihood, 31
Maximum Likelihood estimate, 45
Maximum Likelihood estimator, 31, 50,
    57, 84
Metropolis-Hastings, 102
minimization problem, 32, 34, 36, 57,
    70, 72–74, 76, 109, 116, 193

minimum norm solution, 62, 67
mixed central moment, 9
model reduction, 120, 128
moment, kth, 8
Monte Carlo integration, 92
Morozov discrepancy principle, 81, 85
multivariate Gaussian random variable, 35
multivariate random variable, 11

noise reduction, 28
noise, additive, 42
noise, multiplicative, 45
noise, Poisson, 45
non-parametric, 21, 50
normal distribution, 16
normal equations, 64, 72, 85, 87
normal random variable, 16
nuisance parameter, 41
numerical differentiation, 85

odds, 3
off-centered covariance, 126
orthogonal random variables, 10
outlier, 121
overdetermined problem, 61, 150
overregularization, 87

parametric, 21, 50
Paris, 39
penalty term, 58, 110
pointwise predictive output envelope, 140, 158
Poisson density, 17, 33, 45
Poisson distribution, 102
Poisson process, 14, 44
positive definite, 28, 64
positive semi-definite, 12
posterior density, 55
Posterior Mean estimate, 57
preconditioner, 107
preconditioner, left (right), 108
preconditioning, 77
preprior, 152
Principal Component Analysis (PCA), 118, 128
prior distribution, 51
prior information, 40
priorconditioner, 108, 194

probability density, 6, 7
probability distribution, 7
probability measure, 6
probability space, 6
proposal distribution, 99, 104, 170
pseudoinverse, 65, 67, 76, 78

QR factorization, 65
qualitative vs. quantitative information, 50

random variable, 2, 7
random walk, 161
Rayleigh distribution, 19, 37
reality, 41
regularization, 57, 80
regularization by truncated iteration, 80
regularization parameter, 85
rejection sampling algorithm, 103
residual error of the normal equations, 73

sample covariance, 28
sample history, 176
sampling, 91
Schur complement, 131, 133, 148
search direction, 71
semiconvergence, 79, 83
sensitivity, 80
singular value decomposition, 65, 77, 85, 87, 119
smoothness prior, first order, 115, 183, 188
smoothness prior, second order, 54, 141, 160
standard normal random variable or distribution, 16
statistical inference, 2, 21
stopping rule, 80
structural prior, 94
subjective probability, 1, 2
surrogate distribution, 102

Tikhonov functional, 57
Tikhonov regularization, 58, 85, 109, 116, 150
time invariant kernel, 166
Toeplitz matrix, 44
training set, 118

transition kernel, 166
transition matrix, 162
transition probability, 162
truncation index, 116

uncorrelated random variables, 10
underdetermined problem, 61, 114, 150
underregularization, 87

variance, 8
variance, Poisson process, 15

weighted least squares problem, 37
white noise, 93
whitened PCA model, 121
whitening, 92
whitening matrix, 93
Wiener filtering, 150